Grant Allen

Force and Energy

A theory of dynamics

Grant Allen

Force and Energy
A theory of dynamics

ISBN/EAN: 9783337776169

Printed in Europe, USA, Canada, Australia, Japan

Cover: Foto ©berggeist007 / pixelio.de

More available books at **www.hansebooks.com**

FORCE AND ENERGY

A THEORY OF DYNAMICS

BY

GRANT ALLEN

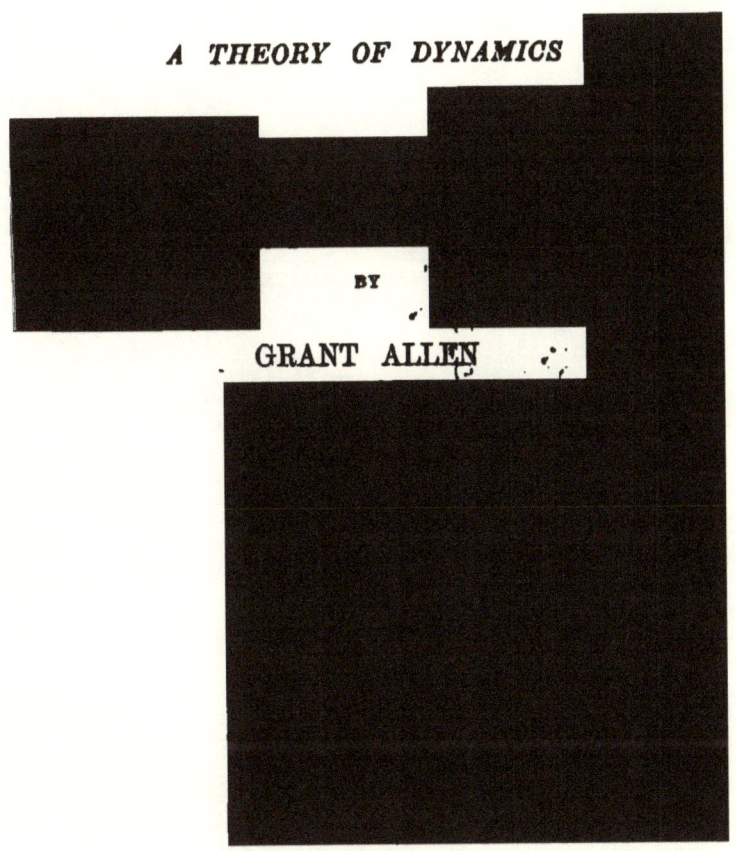

LONDON

LONGMANS, GREEN, AND CO

AND NEW YORK: 15 EAST 16th STREET

1888

POSTERIS

INVENTIONIS FORTASSE PIGNVS

APOLOGY.

Iᴛ is with the profoundest diffidence that I set forth
this book. My best excuse for its publication now may
probably be found in the circumstances under which
I have been induced at last to rush into print with it.
The work has lain by me for nearly double the time
prescribed in the familiar Horatian maxim. Some
fourteen years ago, when I was head of a Government
college in Jamaica, the perusal of certain dynamical
treatises of Clerk Maxwell's, Tait's, Balfour Stewart's,
and Helmholtz's, suggested to my mind sundry pro-
found difficulties in the current conception of the
nature of Energy. Puzzling out these difficulties
conscientiously with myself, as best I might, I began
at length to see, or think I saw, a way out of them
by means of a new theory of my own. This theory,
which, right or wrong, gradually grew clear to my
mental vision, I embodied in a little twenty-page
pamphlet bearing the same title as the present work,
and printed privately at Oxford in 1875 for distribu-

tion to a few physical specialists. Not many of the specialists, I fear, looked at my lucubrations : those who did returned me one or other of two apparently contradictory criticisms. Some of them said my theory was only just what was already known and universally acknowledged. .Others of them said it was diametrically opposed to what was already known, and betrayed an elementary ignorance of the entire matter. To the ignorance thus imputed I will candidly plead guilty, and will proceed to explain why, in spite of it, I have ventured after so long a lapse of time to obtrude my speculations upon a learned audience.

In 1877 I returned once more definitely to the subject, in which my interest had never in any way declined, and, mainly for the sake of clarifying and systematising my own conceptions, worked out my nebulous ideas in full in the present treatise. But finding from the reception accorded to my tentative little pamphlet that physicists were not likely (then, if ever) to admit my contention, and convinced that they knew a great deal more about the matter at stake than I did, I put the completed manuscript severely away in my desk, where it has remained ever since in peace and quiet among a great many more rejected juvenile performances. There it might have remained to all time but for an accidental coincidence which happened a few years back.

The coincidence came about in this way. My friend Edward Clodd submitted to me in the summer of 1885 the first rough sketch of his recent work ' The Story of Creation.' In discussing with him the outline of that book, and especially certain points connected with his conception of Force as there embodied, I found he had lighted upon some of the self-same fundamental difficulties which had originally led me to the views set forth in this little volume. In the course of our conversations on these moot questions I ventured very gently to hint at my own heresies, while disclaiming any desire to poison his mind with them : indeed, so anxious was I not to mislead my friend in this matter that it was with great reluctance I at last consented to lend him the old and crumpled manuscript of my early essay. On reading it over, he told me it had entirely dissipated his difficulties, and had set the whole question for him in a new light. Furthermore, to my unfeigned dismay and distress, he announced that he intended to embody the theory in outline in the dynamical portion of his forthcoming work. Much alarmed, I endeavoured to dissuade him from so rash a course, seeing that like myself he was no physicist, and that the doctrine was new, strange, and heterodox : but so great was his confidence in the truth of the theory that my protests fell flat upon unwilling ears. He incorporated the heretical conception in ‘ The Story of Creation,’ and, as I feared beforehand, suffered not a little for his generous rashness at the hands of the critics.

A monomaniac who has found one other person to share his monomania might perhaps have been excused for jumping to the conclusion that the rest of the world would probably give him a fair hearing. But I was far too afraid of mathematical opinion to venture even so upon publishing my probably crude and incorrect ideas. I still refrained from any attempt to print my book, till I saw that the attacks upon Mr. Clodd’s position almost made it a point of honour for me to lay the facts in their integrity before the judgment of the scientific world. It was not right my friend should suffer for my own transgression. Criticism was levelled at the necessarily brief and bald abstract he had given of what I may venture to call our joint opinion : I thought it only proper, in justice to him, that the theory as a whole should be put in evidence for the jury of experts to examine and decide upon. I don’t for a moment suppose they will take the trouble to look into it at all : but at any rate I have now discharged my duty—*liberavi animam meam*—the evidence is here, and who will may consider it.

Nobody could be more sensible than I am how little likely it is that a mere amateur should hit upon

a true generalisation in science missed by the recog-
nised leaders of physical thought. For this reason, I
would never have published my treatise at all (pro-
foundly as I myself believe in it) had it not been for
Mr. Clodd's intervention, with its remoter consequences.
As it is, however, I may plead in extenuation this
further excuse. The thoughts one entertains, says the
greatest of living English thinkers, are as children born
to one which one may not willingly let die. There
can be no harm, therefore, in putting them forth to
the world, in a tentative way, with all due modesty,
provided always it is clearly understood that they are
put forth as suggestions alone, for wiser heads to accept or reject at leisure. If perchance it should
happen that one has indeed hit almost by accident
upon a true and luminous principle, one owes it to
humanity to set that principle forward at once, in
spite of the natural fear of criticism and ridicule.
The would-be discoverer is probably wrong: but
when by any stroke of luck he chances to be right, it
is for the good of the world that he should publish
his discovery. In this light, therefore, I venture to
beg the professional critic to examine my work. It
pretends to be no more than a suggestion, an *aperçu*,
an attempt at a theory : I ask for it nothing better
than honest consideration : for if this counsel or this
work be of men, it will come to nought : and I have

no desire to aid in the promulgation or diffusion of error.

For the same reason, I will not apologise for the seemingly dogmatic mould in which the treatise itself is cast. Being nothing more than an endeavour to express in words the fundamental dynamical constitution of the universe, as it envisages itself to a particular inquirer, I have thought it best to use the purely impersonal form, and to state each proposition as simple fact, leaving the reader to bear in mind for himself throughout, that the whole is suggestion or conception merely.

At the same time, I sincerely trust scientific readers (if I am fortunate enough to attract any) will approach the theory with an unbiassed mind, and instead of rejecting it offhand at the first glance, because its conceptions do not agree with those to which they are already accustomed, will do me the justice to read it through before deciding, and to place themselves as far as possible in sympathy with my point of view. I grant at once that the idea of Energy they will here find embodied is not at all the idea hitherto framed by men of science. It is a new idea ; and that is exactly why I have written this little treatise. If I am right (as I probably am not) our concepts of Energy will have to undergo a considerable revision. That being just the question at issue here, I hope readers will duly consider it, instead of taking the current view dogmatically for granted, and crushing me by pointing out that mine does not coincide with it. A *petitio principii* is no refutation.

The long time I have kept this treatise by me unpublished ought to supply sufficient proof of the extreme timidity with which I myself regard it. That timidity may perhaps be allowed to protect me from harsh, unkindly, and contemptuous criticism. If I am wrong, of course, I shall expect to be frankly told so : I shall accept demonstration of my mistakes and misconceptions with a good grace. Naturally, I shall continue still to think myself right : it is not in human nature to do otherwise : the theory has too long interwoven itself into all my conceptions of the physical world to be easily rooted out of the fibres of my brain now after so many years. But having once consented to trot out my little heresy unwillingly before the eyes of the world, I shall drop it in public henceforth and for ever. I will make no angry replies to authoritative expositions of my blunders or errors : I will abstain from imitating the common paradox-monger, who, hardened in his obliquity, sees only unfair attacks and unworthy motives in demonstrative criticism. ' I'm not a-arguin' with you ; I'm only a-tellin' of you,' said a pothouse politician to an

obtuse friend. I don't expect to be argued with : I
shall be satisfied to be told.

Under these circumstances, and in consideration
of previous good conduct, I earnestly trust the court
of scientific opinion will let me off with a caution or
a nominal fine. My promise never to recur to the
subject again may surely in such a case be counted
to me for righteousness. At a certain college examination, where proof of age was required from all
intending candidates, a certain colonial-born undergraduate brought with him perforce his only docu-
mentary evidence, a certificate of baptism. The
examiner, a well-known heterodox don, glanced at the
ecclesiastical certificate curiously. 'How's ·this ?'
he asked in a hurried voice. 'How's this? You've
been baptised, sir?' The luckless undergraduate
timidly stammered out that it was a mistake due to
the imperfect registration system of his native land.
'H'm,' snorted the examiner : ' oh, very well, then :
as you were baptised by mistake, it won't be allowed
to tell against you.' May I venture to express a
humble hope that on this occasion too a heresy
extorted from me under such peculiar circumstances
will not be allowed to tell against my character ?

CONTENTS.

PART I.—ABSTRACT OR ANALYTIC.

PART II.—CONCRETE OR SYNTHETIC.

ABSTRACT OR ANALYTIC

PART II.—CONCRETE OR SYNTHETIC.

Part I.

ABSTRACT OR ANALYTIC

B

CHAPTER I.

POWER.

A POWER is that which initiates or terminates, accelerates or retards, motion in one or more particles of ponderable matter or of the ethereal medium.

Power, as here understood, is thus the widest of all possible dynamical conceptions. It cannot be defined by genus and differentia, because it is itself the *summum genus* of dynamical science. Accordingly, it will be observed that no attempt is made above to assign it to any higher class, such as *things*, *entities*, or *concepts*. Nothing would be gained, for example, by saying that a power is the *tendency* to initiate or terminate motion : it is best described by the indefinite statement given at the head of this chapter. It is simply *that which produces or destroys*, *increases or lessens*, motion in any particle or particles of any substance whatsoever cognisable by man.

Powers are of two sorts, Forces and Energies, the differences between which will be fully set forth in subsequent chapters. Meanwhile, as a help to the provisional comprehension of the nature of Power, which can scarcely be grasped at first in the abstract terms of our formal definition, it may be mentioned that amongst the varieties of Power are such Forces as Gravitation, Cohesion, and Chemical Affinity, besides such Energies as Heat, Electricity, and Light. These expressions are here employed in their popular sense, merely as guides to the sort of concept provisionally set forward for the term Power, until the subsequent investigation has rendered possible a more rational and comprehensive notion in the mind of the reader.

CHAPTER II.

FORCE.

A FORCE is a Power which initiates or accelerates aggregative motion, while it resists or retards separative motion, in two or more particles of ponderable matter (and possibly also of the ethereal medium).

All particles possess the Power of attracting one another—in other words, of setting up mutually aggregative motion—unless prevented by some other Power of an opposite nature. Thus a body suspended freely in the air is attracted towards the earth by the Force (or aggregative Power) known as Gravitation. A piece of sugar, held close over a cup of tea, attracts into itself the water of the tea-cup, by the Force (or aggregative Power) known as Capillarity. A spoon left in tea grounds or a foot planted on the moist sand similarly attracts the neighbouring drops. A piece of iron or coal exposed to free oxygen (each at a certain fixed temperature) attracts the particles of oxygen by the Force known as Chemical Affinity. In every case there must be an absence of counteracting

Energies (or separative Powers) sufficient to prevent the union of the particles, as will be shown hereafter : but for the present it will be enough to notice that every particle attracts every other particle in some one of various ways, unless prevented . by other Powers.[1]

Not only, however, do all particles thus attract one another, but they also resist all attempts to separate them from one another. A weight suspended in the air falls to the ground : but it also resists any attempt to remove it from the ground, which can only be done by the employment of a proportionate Energy (or separative Power). The water which the sugar has absorbed can only be drawn from it by the Energy of suction. The oxygen with which the iron has united can only be driven off by the Energy of heat : while the carbonic anhydride and water which resulted from the burning of the coal yield only as a rule to the separative Energy of light or electricity. In every case the Force which brought two or more particles together in the first instance keeps them united ever after, and must be neutralised by an equal Power of an opposite description before they can be disjoined.

[1] The term 'to attract' must be strictly accepted in the sense of actually setting up aggregative motion, not merely in that of a tendency to such motion. The tendency always subsists, in spite of counteracting causes, and is immediately actualised upon their removal. -

CHAPTER III.

ENERGY.

AN Energy is a power which resists or retards aggregative motion, while it initiates or accelerates separative motion, in two or more particles of ponderable matter or of the ethereal medium.

All particles, or aggregates of particles, not actually in contact with one another in stable equilibrium at the absolute zero of temperature, are kept apart by an Energy or separative Power of some sort, which prevents them from aggregating as they would otherwise do under the influence of the Forces inherent in them. Thus the moon is prevented from falling upon the earth, and the earth from falling into the sun, by the Energy of their respective orbital motions. A ball shot from a cannon into the air is prevented from falling by the Energy of its upward flight. A red-hot poker has its particles kept apart by the Energy of heat. In every case, so soon as the Energy is dissipated (as hereafter explained) the ball yields to

the aggregative Power of Gravitation, and the poker
contracts to its ordinary dimensions; while there is
no reason to doubt that under similar circumstances
the moon and the earth will aggregate with the sun.
The particles of water are kept in the liquid state by
the Energy known as *latent heat,*[1] and so are those
of steam : when the 'latent heat' is dissipated, the
steam condenses and the water freezes.   There are
many apparent exceptions ; but they will be considered at later stages of the argument.   For the
present, the reader must be content to understand
the word *Energy* (when used in this treatise) only in
the sense here given to it of a Power which resists
or retards aggregation.

Energies also initiate separative motions.   Thus,
a cannon ball is raised by Energy to a distance from
the main mass of the earth which usually holds it
bound by Gravitation on its surface.   A poker placed
in the fire has its particles separated from one another
by the Energy of Heat.   When ice melts or water is
converted into steam, the same Energy similarly severs
their particles from one another and places them in
positions of relative freedom.   In the electrolysis of
water the Energy of the galvanic current tears asunder
the atoms of hydrogen and oxygen from their close

---

[1] I continue to employ for the present this well-known but very
incorrect expression.

union in the compound molecule. In short, wherever we see masses or particles in the act of separating from one another, we know that the separation is due to some Energy.

CHAPTER IV.

THE SPECIES OF FORCE.

FORCES may be most conveniently divided according to the nature of the particles or bodies in which they initiate and accelerate aggregative motion or resist and retard separative motion. Of these, there are four principal kinds known to us or conjectured by us. The first kind is the Mass or visible aggregation of particles, which admits of mechanical separation into minor masses. The second kind is the Molecule, or ultimate mechanical unit, which does not admit of subdivision, except by resolution into its chemical components. The third kind is the Atom, or ultimate chemical unit, which does not admit of subdivision by any known means, though it may perhaps be resoluble hereafter into some simpler and more primitive units. The fourth is the Electrical Unit,[1] whose nature is very inadequately known to us, but which must be considered for our present purpose as in some way

---

[1] This conception of electrical units is provisional and purely symbolical; but its use will be apparent in later chapters.

the analogue of the others, though we have no sufficient warrant for giving it any material properties.

The Force which aggregates Masses and resists the
separation of Masses is known as GRAVITATION. When
any two Masses are left free to act upon one another
without the counteracting influence of an Energy,
they aggregate in obedience to this Power. When
the cannon ball falls upon the earth, it is Gravitation
which draws them together. When an aërolite comes
within the circle of the earth's attraction, it is Gravitation which makes them leap towards one another.
If the moon were to lose its orbital Energy, Gravitation would pull it to the earth ; and if our planet in
her turn were suddenly checked in her course, Gravitation would cause her to plunge into the sun, while
the sun in return would make a slight bound to meet
her.   Again, when any two Masses are in a state
of aggregation, the Force of Gravitation resists any
attempt to sever them. If the cannon ball lies upon
the ground, it cannot be raised without an expenditure
of Energy, and the amount of the Energy required
to lift it to a given height (or distance from the surface
of the earth) is the measure of the resistance offered
by Gravitation. Similarly, when the Masses are not in
actual contact owing to the existence of an Energy
which keeps them apart, as in the case of the earth
and her satellite, or the sun and the planets, Gravi-

tation resists any attempt to sever them beyond their actual distances. It would be impossible to remove the moon a hundred miles from the earth, or the earth a hundred miles from the sun, except by the employment of an adequate Energy ; and, as in the simpler case, the amount of Energy required would be the measure of resistance offered by Gravitation.

The Force which aggregates Molecules and resists the separation of Molecules is known as COHESION, When any two Molecules are left free to act upon one another without the counteracting influence of an Energy, they aggregate in obedience to this Power. But the cases are much more difficult to illustrate than those of gravitation, because while masses attract one another powerfully at very conspicuous distances, Molecules (practically speaking) only attract one another at infinitesimal distances. The difference, however, which is purely relative, may thus be illustrated and explained. An aërolite is not drawn on to the earth unless it approaches the earth very closely, because otherwise the earth's attraction, though causing a deviation in its course, does not suffice to overcome the aërolite's energy and the combined attractions of surrounding bodies. But if it be near enough to be more powerful than all of them put together,[1]

[1] I take for granted on the reader's part a knowledge of the law of inverse squares.

the aërolite either circles round the earth as a satellite or even falls at once upon its surface. Similarly with Cohesion. If two pieces of uneven iron be laid upon one another, the molecules do not approach near enough to exert any conspicuous mutual influence: but if the two pieces be planed to an absolute smoothness, so that the several molecules can come within the sphere of their mutual attraction, they will cohere perfectly, and it will be impossible to tear them asunder. Again, in other cases, Cohesion can only be effected by such a molecular motion (or heat) as will cause the Molecules to approach one another closer than they can be induced to do by mechanical means: just as an aërolite which would not under ordinary circumstances come (practically speaking) within the sphere of the earth's attraction, might do so if it were given an oscillating motion from side to side, so as to cross or closely approach some portion of the earth's orbit. Thus, two pieces of iron, if heated, will cohere with one another. Furthermore, the molecular motion inherent in the liquid form is often sufficient for this purpose: thus, two masses of dough, which will not cohere in the dry condition, can be made to do so by the addition of moisture. In the practice of gumming and glueing, we make use of this device in everyday life. A further account of these phenomena will be given in the chapter on Liberating Energies. The second property of Cohesion, that of resisting the separation of Molecules actually aggregated, is much more familiar to us. If two Molecules or bodies of Molecules are in an aggregated condition—that is, are not rendered plastic or liquid or gaseous by some form of Energy—we cannot separate them without a considerable expenditure of Energy. The Energy may be in the form of a mechanical action, as when we tear or break a cohering substance; or of heat, as when we melt lead; or of the contained motion of liquids, as when we dissolve a lump of sugar. But in any case Energy must be expended to counteract the aggregative Force of Cohesion in solid bodies.

A qualification must be added to prevent misconception. The cohering Molecules need not be supposed to be in actual physical contact with one another. It is sufficient that they should be within the sphere of one another's attraction; just as the moon is kept in its place by the earth, and the planets by the sun, in spite of the intervening space. Theoretically, of course, every body in the universe attracts every other; but as the attraction decreases as the squares of the distance, at practically infinite distances it becomes practically infinitesimal and can be overcome by an infinitesimal Energy. This is the case ordinarily with Cohesion: at very slight distances its Force is so diminished that only an imperceptible amount of Energy is required to counteract it. But there is no reason to doubt that when the two rough pieces of iron are laid upon one another, the supporting points, so to speak, come within the sphere of mutual attraction, though their number and area are so small that we cannot perceive the resistance resulting from their Cohesion when we separate the pieces. In short, Cohesion always tends to act between all Molecules, but its effects may be disguised either by distance or by counteracting Energies. Other cases will be treated in the chapter on Mutual Interference of Forces. Adhesion and Capillarity are only forms of Cohesion.

The Force which aggregates Atoms and resists the separation of Atoms is known as CHEMICAL AFFINITY. As here employed it will be understood to mean not merely the Force which unites the Atoms of two or more elements into a compound molecule, but also the identical Force which unites two or more Atoms of the same element into a molecule such as that of ozone. When any two or more Atoms (or equivalents in combining proportions) are left free to act upon one another without the counteracting influence of an Energy, they aggregate in obedience to this Power. As in the case of cohesion, however, the Atoms must be brought into close contact with one another. When phosphorus is exposed to oxygen the aggregation is immediate. But in other cases a certain amount of molecular or Atomic motion is needed in order to bring the Atoms within the sphere of their mutual attractions. Thus heat is necessary to make carbon combine with oxygen, as in the ordinary phenomenon of combustion : while the more subtle motion of light suffices to effect a union between hydrogen and chlorine. But we may broadly assert that whenever free Atoms find themselves in the presence of a free Atom for which they have affinities (the proper proportions being of course supposed), and are brought within the sphere of their mutual attraction, the two Atoms or sets of Atoms aggregate under the influence of Chemical Attraction.   Here, again, a qualification is needed.   The above rule holds only for *free* Atoms. Just as a ball suspended by a rope from the ceiling does not fall to the ground, because the Force of cohesion outbalances the Force of gravitation, so, when two or more Atoms, united in stable combination, are brought into contact with other Atoms for which they have affinities less strong than those of their existing combination, they will not yield up their stronger to their weaker affinity.   (See the subsequent chapter on Mutual Interference of Forces.)   And again, just as the ball will break the rope, if gravitation outbalances cohesion; so, if the new affinities are stronger than the old ones, the Atoms will yield up their previous combination and enter into that to which they are most powerfully attracted. The second mode in which Chemical Affinity acts is in resisting the attempt to separate the component Atoms of a compound body. Setting aside for the present certain very abnormal cases in which 'unstable' bodies spontaneously decompose—cases which can only be explained at a very late stage of our exposition—all ordinary 'stable' compounds require an Energy to separate their Atoms. Thus heat is needed to divide the Atoms of oxygen from those of iron in ferric oxide : while electricity is necessary to sever the Atoms of hydrogen from those of oxygen in water. This statement must be understood as applying only to the separation of *free* elements, not the formation of new compounds. Mere juxtaposition is sufficient to make certain compound bodies yield up their weaker affinities in the presence of stronger ones: but (with the special exception noted above, chiefly referring to organic compounds) an Energy is required to separate any compound into its component Atoms in a free state, without the aid of stronger antagonistic affinities.

The Force which aggregates Electrical Units and resists the separation of Electrical Units is known as

c

ELECTRICAL AFFINITY. This Force is little understood, and can only be treated in a very symbolical manner. What few points can be formulated are briefly these. When Positive and Negative Electricities are left free to act within the sphere of their mutual attractions, they are aggregated by this Force, as in the discharge of a Leyden jar. In saying this, no implication of materiality is meant to be conveyed. In our present ignorance on the subject, Electrical Affinity must be placed in the same category as other Forces; though further researches will doubtless enable us to give a better account of its real nature. Similarly, an Energy is necessary to separate the Positive and Negative Electricities which subsist in combination in every material body. In the case of a glass rod or an electrical machine this Energy is that of mechanical motion: in certain other cases it is of thermal or chemical origin. These points will receive further consideration in the chapter on Electrical Phenomena.

A table will put in a clearer light the classification here adopted.

FORCES OR AGGREGATIVE POWERS.

Molar	Molecular	Atomic	Electric
Gravitation	Cohesion	Chemical Affinity	Electrical Affinity

CHAPTER V.

THE SPECIES OF ENERGY.

ENERGIES may be most conveniently divided on the same principle as Forces, according to the nature of the particles of bodies in which they initiate or accelerate separative motion, and resist or retard aggregative motion. But owing to the existence of two modes of Energy, the Potential and the Kinetic, whose peculiarities will form the subject of our next chapter, it will not be possible to assign a single definite name to each species, as was the case with the various Forces. It must suffice for the present to quote a few well-known instances of each.

The energies which separate Masses and resist the aggregation of Masses may be summed up under the title of MOLAR ENERGIES.[1]   Of Molar Energies employed in actual separation, a familiar instance is

---

[1] We shall see hereafter that these species are in reality just as simple as those of Forces; but for the reader's convenience they are exhibited here under familiar aspects, which give them an appearance of plurality and indefiniteness.

c 2

given in our own persons, when we lift a weight from
the ground or carry ourselves to the top of a hill,
thereby counteracting the Molar Force of gravitation
by raising a body to a greater distance than before
from the centre of the earth's attraction. Another
instance is seen in a cannon ball fired vertically, or
a stone lifted by a crane. On a larger scale, any
fresh Energy employed in removing the moon further
from the earth or a planet from the sun would
be a Molar Energy. Any Mass thus separated from
another attracting Mass is said in the current language
of physics to possess Visible Energy of Position, a term
which we shall examine and endeavour to amend
hereafter.   Of Molar Energies employed in resistance to aggregation the most familiar instance is that
of orbital movement. The moon is prevented by this
Energy from aggregating with the earth, and the
planets with the sun, as they would otherwise do
under the influence of Molar Force or gravitation.
On a smaller scale, the Energy of a bird in flying or
a cannon ball fired horizontally is largely employed
in counteracting gravitation. It is seldom, however,
that we see Energy thus employed, except in the
case of the heavenly bodies, because the Molar Force
exerted by the earth in its immediate vicinity is so
strong as to overcome ordinary Energies after a very
short period of dissipation. Masses of the sort here

described are said in the current phraseology to
possess Energy of Visible Motion, which expression,
like the former one, will receive attention at a later
point.

- The Energies which separate Molecules and resist
the aggregation of Molecules may be summed up
under the title of MOLECULAR ENERGIES. Of Molecular Energies employed in separation we have a
common instance in heat, which draws apart the
Molecules of a red-hot poker or a mass of boiling
water, in opposition to the Molecular Force of Cohesion. The contained Energy of water acts in the same
manner on a lump of sugar or a mass of dry dough.
Of Molecular Energies employed in resistance to
aggregation, heat under its converse aspect affords
us an example. The Molecules of all bodies are
prevented from aggregating into their most compressed form by the presence of heat. Thus the red-
hot poker only contracts so fast as it loses its Energy
by radiation. The contained Energy (or 'latent
heat ') of water similarly prevents its aggregation into
ice. Large masses of water before freezing part with
their Energy in the visible form of heated mist.

The Energies which separate Atoms and resist the
aggregation of Atoms may be summed up under the
title of CHEMICAL ENERGIES. A caution as to the
sense in which this term must be here accepted is

appended below.   Of Chemical Energies employed
in separation we have an instance in the electrolysis
of water. The Energy disengaged by the union of
elements in the battery is used up in producing
chemical separation between the atoms of the electro-
lyte. Light produces a similar effect upon carbonic
anhydride and water in the leaves of plants. Any
Energy which separates a compound body into simpler
or elementary bodies may be regarded as a Chemical
Energy in the sense here intended.   Of Chemical
Energies employed in resistance to aggregation, no
unequivocal instance can be cited at our present
stage, though this apparent anomaly will be cleared
up as we proceed. For the time the reader must be
content to accept as an instance the fact that many
Atoms will not combine with one another at a certain
high temperature : the same temperature, in fact, at
which they are driven off from their combination
when actual.   It will be noticed that, for the sake
of uniformity, a somewhat new sense has here been
given to the term 'Chemical Energy.' As ordinarily
used at present, that term refers to the strength of
the tendency which a body shows to unite with other
bodies. It will be seen in the sequel that this is
really a property depending upon separation and
chemical nature : just as a body in proportion to its
height and mass shows a tendency to aggregate with

the earth : but, meanwhile, it is necessary to impose a new sense upon the term, in keeping with the analogous term 'Chemical Affinity,' which we have applied to the Force that aggregates Atoms.

The Energies which separate Electrical Units and resist the aggregation of Electrical Units may be summed up under the title of ELECTRICAL ENERGIES. As in the case of Electrical Forces, our treatment of this department must be considered purely temporary and symbolical. Of Electrical Energies employed in separation we have an instance in the electrical machine, where friction produces a disunion of the Positive and Negative Units. Similarly in the torpedo and gymnotus. Of Electrical Energies employed in resisting aggregation there is again no unequivocal instance. The illustration of this deficiency must be left to later chapters.

Throughout, both in the case of Forces and Energies, it will be noticed that the same Power which initiates and accelerates one kind of motion equally resists and retards the other kind of motion. Thus, Gravitation both initiates movements of masses towards centres of attraction, and resists movements of masses away from centres of attraction. Cohesion both draws molecules together, and resists the separation of molecules : while heat draws molecules apart and resists the aggregation of molecules. So that these two Powers, the aggregative and the separative, are incessantly opposing and antagonising one another in all bodies, great or small. The amount of aggregation reached by any system of bodies at any point of time depends upon the relative proportions of its Forces and its Energies at that moment.

A table is scarcely needed for the contents of this chapter; yet for the sake of symmetry one is here appended.

ENERGIES OR SEPARATIVE POWERS.

Molar	Molecular	Atomic	Electric
Molar Energy	Molecular Energy	Chemical Energy	Electrical Energy

CHAPTER VI.

THE MODES OF ENERGY.

ENERGY has two Modes, ordinarily known as the Potential and the Kinetic : but the terms Statical and Dynamical are much preferable. Nevertheless, in order not to disturb unnecessarily the received terminology, the former expressions will be generally preserved in this treatise.

The two Modes of Energy are interchangeable with one another : the Potential can pass into the Kinetic, and the Kinetic into the Potential. Each species of Energy, Molar, Molecular, Atomic, and Electrical, is represented in both modes.

POTENTIAL ENERGY (a very bad name) is equivalent to actual or statical separation. Any mass, molecule, atom, or electrical unit, in state of separation from other masses, molecules, atoms, or electrical units, possesses Potential Energy. The subject may conveniently be considered under the four heads hence arising.

Molar Potential Energy is equivalent to the statical separation of Masses. The moon possesses this Energy relatively to the earth, and the planets to the sun.[1] The cannon ball, shot vertically, has Molar Potential Energy at the instantaneous neutral point when it has reached its greatest height and has not yet begun to fall. A stone on a mountain top or a head of water on its side has also the same Energy. In short, Molar Potential Energy is possessed by all discrete Masses in virtue of their separation. It is commonly known as Visible Energy of Position.

Molecular Potential Energy is equivalent to the statical separation of Molecules. Two planed surfaces of iron possess this Energy, until by apposition they are made to unite. The molecules of water, dispersed as steam, similarly possess it, in the form commonly known as 'latent heat.' When steam condenses or water freezes, the Energy is yielded up in the Kinetic form.

Atomic Potential Energy is equivalent to the statical separation of Atoms. It is possessed by every free Atom of an element, and by every compound Atom whose affinities are not fully saturated.

---

[1] It may excite surprise to see these relations described as statical. The term is only employed in a relative sense, as opposed to the dynamical energy of a falling body.

Thus an Atom of carbon has Potential Energy in relation to two separate Atoms of oxygen, with which it may unite to form carbonic anhydride. Similarly, chlorine has Potential Energy relatively to sodium, with which it may unite to form common salt. Such cases, however, must be carefully distinguished from those of preferential attraction where a body leaves its union with one element to combine with another for which it has stronger affinities : as when the Cl of HCl leaves the H to unite with Na in NaCl. This last instance is really analogous to that of the cannon ball which breaks the rope that ties it because the Force of Gravitation has outbalanced that of Cohesion.

Electrical Potential Energy is equivalent to the statical separation of Electrical Units. In a Leyden jar, the opposite electricities of the inner and outer coats exhibit this relation. In a thunder cloud and the earth beneath it we have a substantially similar division of the Positive and Negative Units. The statement of these facts must be accepted with the usual caution as to the purely symbolical nature of our electrical conceptions.

From the potential we pass on to the Kinetic Mode. It will not be immediately apparent in what sense Kinesis is an Energy in accordance with our definition : but, here again, the reader must courteously waive his objections for the present, and accept the statement provisionally, so far as he finds possible. Many difficulties of this sort necessarily beset the explanation of every new point of view, especially where previous misconceptions have clouded and embarrassed the mental vision.

KINETIC ENERGY is equivalent to motion. Any mass, molecule, atom, or electrical unit, in a state of motion, possesses Kinetic Energy. The subject may be conveniently considered under the four heads hence arising. But, just as before, when dealing with Energy generally, we found that we could not divide it into species so definite in their likeness as those of Force, because Energy was manifested in two Modes, the Potential and the Kinetic: so, here, when we are dealing with Kinetic Energy specially, we shall find that it cannot be divided into species so definite as those of the Potential Mode, because Kinesis itself is divisible into several Kinds, whose nature will form the subject-matter of the succeeding chapter.

Molar Kinetic Energy is equivalent to the relative motion of Masses. It is seen in the fall of an unsupported weight or a spent cannon ball to the earth. It is also seen in the rising of the ball, the flying of a bird, or the walk of a man. Again, it is seen in the orbital motion of the planets, and in the spinning of a top.

These various Kinds of Kinesis will be fully discussed in the next chapter.

Molecular Kinetic Energy is equivalent to the relative motion of Molecules. It is found in the falling together of Molecules of steam into water. It also occurs in the disruption of a cohering mass. And it is more conspicuous in the phenomenon of heat.

Atomic Kinetic Energy is equivalent to the relative motion of Atoms. It is seen in that rushing together of Atoms which results in chemical combination. It also occurs in the severing of Atoms from the combined state. But it is not known to have any continuous form analogous to the orbital motion of a planet, the spinning of a top, or the regular vibration of heat.

Electrical Kinetic Energy is equivalent to the relative motion of Electrical Units. It is seen in the lightning, in the discharge of a Leyden jar, and in the galvanic current.

It will doubtless seem strange to the reader to find the motion of masses, molecules, and atoms *towards* one another spoken of as a manifestation of Energy: but this seeming inconsistency will be explained in the succeeding chapter.

A table will clearly exhibit the relations here described, one example only of each species being cited.

ENERGIES OR SEPARATIVE POWERS.

MODES	SPECIES			
Potential	Molar Potential Energy. (Visible Energy of Position.)	Molecular Potential Energy. (Condensing Steam.)	Atomic Potential Energy. (Chemical Energy of Free Elements.)	Electrical Potential Energy. (Tension.)
Kinetic .	Molar Kinetic Energy. (Orbital Motion.)	Molecular Kinetic Energy. (Heat.)	Atomic Kinetic Energy. (Chemical Energy in Act of Combining.)	Electrical Kinetic Energy. (Galvanic Current.)

CHAPTER VII.

THE KINDS OF KINESIS.

MOTION has three Kinds, considered from our present standpoint. It may be separative, or it may be aggregative, or it may be continuous and neutral. Each species of Kinetic Energy has a form of each Kind.

Molar motion may be separative, as when a cannon ball is shot up into the air; or aggregative, as when the same cannon ball falls to the earth; or continuous and neutral, as when a top spins in the same place.

Molecular motion may be separative, as in tearing asunder a mass; or aggregative, as in condensing steam; or continuous and neutral, as in the case of heat.

Atomic motion may be separative, as in decomposition; or aggregative, as in the act of combining. The continuous and neutral stage is not at present known, though there is reason to think that it exists.

Electrical motion may be separative, as when the Positive and Negative Electricities are divided; or aggregative, as when they are uniting. The continuous stage is possibly given us in the current which is supposed to circle round a magnet.

It was noticed in the last chapter that there was an appearance of contradiction in the statement that aggregative motions were yet manifestations of Energy. That difficulty must now be met.

When a cannon ball is shot up into the air, the motion is obviously separative, and there can be no doubt of its being a manifestation of Energy. Similarly, when a set of molecules are separated by mechanical Power or by heat, when a chemical compound is broken up into its elements, and when the Positive and Negative Electricities are sundered from one another, the separative nature of the process is obvious. We can have no hesitation in assigning each of these cases to the action of an Energy.

But when we look at the continuous and neutral motions, their character as Energies is less obvious. A moment's consideration, however, will make it clear. The orbital motion of the planets is a continuous Energy which prevents them from aggregating with the sun as they would otherwise do. The motion of the top in like manner prevents it from falling on to the earth. The continuous vibratory molecular motion (or heat) of the red-hot poker prevents the steam or the water particles from aggregating into their cooled or liquid or solid states respectively. In short, whenever a body or molecule in a free state does not aggregate immediately with the other bodies or molecules which attract it, it is kept apart from them in virtue of some continuous or neutral movement.[1] So soon as it parts with its Energy (or motion), it aggregates with the attracting body. Thus when the steam loses its heat it condenses into water; when the water in turn is deprived of heat, it freezes into ice; when the poker cools, it contracts ; when the top parts with its motion to the air on the surface, it falls ; and we have no reason to doubt that when the planets have dissipated their Energy of orbital movement by ethereal friction they will fall into the sun. This general principle—that *free* bodies can only be kept from aggregating by a continuous movement—is one of great importance, whose value will be seen hereafter. A body in such a state of continuous movement, which prevents it from aggregating with another, is said to be in *equilibrium mobile*.

When, however, we come to the aggregative motions, it would seem at first sight as though these

---

[1] The reader must be cautioned to notice the expression 'in a free state,' which excludes such instances as those of a weight tied by a string, or a chemical body already in stable combination, whose case will be considered in the chapter on the Mutual Interference of Forces.

D

must be classed with Forces, not with Energies. A considerable faculty of abstract thought is required to grasp their real relations : nevertheless we must endeavour to solve the problem. In doing so, we must trench a little on the subject-matter of future chapters, but only by alluding to facts already familiar to the reader. When the cannon ball reaches its highest point it possesses Potential Energy. But it does not remain suspended in the air. There are only two conditions under which it could do so, in opposition to the Force of gravitation : the first is if it is supported by a ledge or rope, in which case cohesion balances gravitation ; the second is if it possesses continuous kinetic energy, in which case it would circle round the earth as a satellite until its energy was dissipated. Practically, the existence of the atmosphere makes the second case purely imaginary within the limits of that. medium, though it is exhibited in the ether by such a body as the moon. As the cannon ball does not fulfil either of these conditions, it begins at once to fall. But the Potential Energy which it possesses becomes thereupon Kinetic, from moment to moment, until, at the instant of touching the earth, it has all assumed that mode. Now, we know that it does not then utterly disappear. The great principle of the Conservation of Energy teaches us that it is changed into the form of heat. Accordingly, while the two masses aggregate, certain molecules of each are separated by heat. At the moment of contact, all the motion of the fall, or Aggregative Molar Kinetic Energy, is changed into heat (or separative Molecular Kinetic Energy). There is just as much separation at last as at first : only when the ball was at its height, the separation was molar ; and when the ball has touched the earth, the separation is molecular. The formula which tells us how many heat-units are generated by the fall of such and such a mass through so many feet, is a formula for the equivalence of molar separation with molecular separation. But in the intermediate time, during the fall, Potential Energy was disappearing every moment, and motion was taking its place. Though this motion was aggregative, yet, when the ground was reached, it changed into the separation of heat. Accordingly, we are justified in regarding it as essentially a transitory form of separative Power. This will be still clearer if we take such a case as the moon's. That satellite, though attracted by the earth, is yet prevented from aggregating by its orbital movement. It possesses Potential Energy in virtue of its separation, but this does not assume the aggregative Kinetic form on account of the continuous orbital Energy. If, however, we suppose the moon to have lost its orbital movement, still retaining its present position and size, it would at once yield to the earth's attraction, and all its Potential Energy would become Kinetic. When it reached the earth, the shock of its fall would reduce it to a very heated state, and an immense increase in size would result from the separation of its particles. The merely transferential nature of the aggregative motion is here clearly seen.   So too, in the case of molecules. The Potential Energy of steam is given up when it condenses into water ; and the Potential Energy of water when it forms into ice.   Similarly with atoms. When oxygen unites with carbon and hydrogen in a candle, their Energy is yielded up in the form of heat, which produces a separation (or rarefaction) in the neighbouring atoms of the atmosphere.   The same truth is shown in the heat and light evolved during the aggregation of Positive and Negative Electricities.   Throughout we see that aggregative Energy is merely Potential Energy in the course of transformation to another form. While the really aggregative Power of Force is causing these bodies to combine, the Energy of their motion represents for a while their original separateness, and is finally transformed into a similar separateness between other bodies.

A concrete instance will make this clearer. Let us suppose the case of a pulley, with a weight at each end, one suspended in the air at the utmost height of the pulley, and the other slightly lighter, on the ground.   The heavier weight possesses Potential

Energy in virtue of its elevation ; but, if it is free to act, it is drawn down by the aggregative Force of gravitation. In this case, however, all its Energy does not assume the Kinetic Mode as it drops : the greater part of it is used up in elevating the lighter weight to the same height, while the remainder chiefly goes off in the form of friction—that is, heat—that is, molecular separation. There is thus a mere fraction left to be converted into heat when the weight touches the ground; the mass of the Energy still remains Potential in the lighter weight. Here we see that the Energy of a falling body does not consist in its mere downward movement, but rather in that accelerating motion which is capable of being transformed into heat when the masses aggregate. If the motion be infinitely slow, the amount of heat evolved will be infinitesimal. So that the Energy of Kinesis is seen to be a mere transferential mode from one kind of separation to another.   Again, we may look at the similar instance of a clock, driven by a weight. Here the weight possesses Potential Energy, in the same way as in the case of the pulley ; but it has opposed to it, not another weight (that is, gravitation), but friction (that is, cohesion).[1] As gravitation pulls down the

---

[1] Above we used friction in a different sense, as equivalent to heat. This is a necessary ambiguity of our present terminology. From the point of view of the Force involved, friction means the cohesion which must be overcome ; but from the point of view of the Energy employed, friction means the separative power of heat which overcomes

weight through each inch of its course, the Potential
Energy so lost assumes the form of heat, or separative Molecular Motion, in the wheels and bearings.
When the weight reaches the ground, its Energy has
all been used up, and the aggregative movement has
been a real display of Force.

Thus all the kinds of motion are ultimately shown
to be forms of Energy or Separative Power.

KINETIC ENERGIES.

Separative	Separative Molar Motion. (In a body raised from the earth's surface.)	Separative Molecular Motion. (In a body torn apart.)	Separative Atomic Motion. (In chemical decomposition.)	Separative Electrical Motion. (In electrical machine.)
Aggregative	Aggregative Molar Motion. (In a falling body.)	Aggregative Molecular Motion. (In a body cooling.)	Aggregative Atomic Motion. (In chemical combination.)	Aggregative Electrical Motion. (In lightning.)
Continuous	Continuous Molar Motion. (In a top or a planet.)	Continuous Molecular Motion. (In heat.)	Continuous Atomic Motion. (Unknown.)	Continuous Electrical Motion. (In magnet ?)

CHAPTER VIII.

THE PERSISTENCE OF FORCE.

EVERY particle of matter has inherent in it certain Forces of which it can never be deprived. The total amount of Force or Aggregative Power in the universe is thus always a fixed quantity. This principle may be known as the Persistence of Force. It must be carefully distinguished from the opposite principle of the Conservation of Energy, to which the same name has been frequently but most incorrectly applied.

Every mass tends always to attract every other mass, and cannot be deprived of this tendency. The tendency may be masked for awhile by the intervention of other masses, as when a loose stone stands on the top of a wall, or by the presence of an Energy, as when the moon circles round the earth, or a ball is shot from a cannon; but it cannot be got rid of: for as soon as the stone topples over with the wind it falls to the ground at once; as soon as the ball parts with its Energy it similarly falls ; and as soon as the moon has got rid of her motion by ethereal friction, she will aggregate with the earth.

Similarly with molecules, atoms, and electrical units : every one of them when in a free state, unrestrained by interfering Forces, and unacted upon by Separating Energies, rushes at once into a state of aggregation with its fellows.

It is important to notice that Force, unlike Energy, is inherent and indefeasible in every unit of matter. It may be counteracted for awhile by an Energy, but it still remains ready to act so soon as the Energy is dissipated ; it never passes from one unit to another, as we shall see that Energy does. Force, or aggregative Power, is the primary and indefeasible attribute of every material particle.

CHAPTER IX.

THE CONSERVATION OF ENERGY.

THE total amount of Energy, Potential and Kinetic, existing in the universe is always a fixed quantity. It is not, however, like Force, rigidly bound up with the individual particles in which it is from time to time manifested. As we have already seen, it can be transferred from one particle or set of particles to another. For this reason it has been deemed desirable to embody the principle in different language from that which we employed in the somewhat analogous case of Force. While Forces *persist*, Energies are *conserved*. The concrete and practical results of this difference are enormous.

It does not come within the scope of the present work to give a full account of the quantitative relations subsisting between the various species of Energy ; it will be sufficient to trace their equivalence in its broader qualitative aspect. For this purpose we may consider the phenomena of Conservation under three heads: the passage of Energy from the Potential Mode to the Kinetic, the passage of Energy from the Kinetic Mode to the Potential, and the passage of Energy from one species of the Kinetic Mode to another. ·

Potential Energy or relative statical separation [1] has a tendency constantly to pass into the Kinetic Mode, under the influence of Force. Every free body or particle, unless restrained by an antagonistic Force, or kept in separation by a continuous Kinetic Energy, is aggregated at once with other bodies or particles which attract it. A mass poised on a ledge or suspended by a rope is prevented from aggregating with the earth by the Force of cohesion ; but when some external Energy has pushed it off the ledge or severed the rope, its Potential Energy passes at once into the Kinetic Mode, under the influence of gravitation. Two molecules of water vapour are prevented from aggregating under the relatively feeble attraction of cohesion at a distance by their inertia—that is, by the relatively strong cohesion of surrounding or intervening matters (just as a mass

---

[1] By this term is implied a separation which, though perhaps accompanied by actual motion, does not carry the two related bodies further away from one another. Thus, orbital motion in a perfect circle, or the upright spinning of a top, is statical relatively to the centre of gravity of the system; while a fresh energy would be required to carry the related bodies further away from one another.

on the table, though attracted by the earth, is prevented from aggregating by the intervention of the cohering boards)—but when some external Energy brings them within such a distance of one another that the resistances are overcome by their mutual attractions, their Potential Energy becomes Kinetic, and they aggregate with one another.   Two atoms (having affinities for one another) are similarly prevented from aggregating by inertia; but when brought within the sphere of their mutual attraction, their Potential Energy becomes at once Kinetic, and they combine with one another.   So also, two electrical units are prevented from aggregating in the Leyden jar by the electrical neutrality of the glass partition; but when a conducting medium is made to connect them, their Potential Energy passes into the Kinetic Mode and they rush together at once.

Kinetic Energy or motion often passes into the Potential Mode. The Kinetic Energy of actual separation always exhibits this interchange.   A cannon ball fired in the air, the piston of a steam-engine forced up by the expansive Energy of the steam, a weight hauled by a pulley to a height, a man who has climbed a mountain, are all of them instances where Molar Kinetic Energy has become Potential.   The liquid condition of water melted from ice, the diffused state of vapour raised from water, are

instances where Molecular Kinetic Energy has become
Potential.   The free hydrogen and oxygen of an
electrolytic bottle, the iron and oxygen driven from
their combination by heat, are instances where Atomic
Kinetic Energy has become Potential.   The negative
and positive electricities of a Leyden jar, of a thunder-
cloud and the earth, of the knobs of an electrical
machine, are instances where Electrical Kinetic
Energy has become Potential.

Finally, Kinetic Energy often passes from one of
its species to another.   Molar motion passes into
Molecular motion whenever one mass interferes with
the motion of another.   This is true whether the
motion is aggregative, or separative, or continuous.
If a cannon ball be allowed to fall to the earth from
a position of Potential Energy, all the Kinetic Energy
which the mass acquires in its fall passes to the
molecular species when it touches the ground.   If it
be fired into the air, and immediately checked by an
iron target, the same result occurs.   And if a top be
stopped in spinning or the moon checked in her
course, exactly like effects are or would be produced.   Molecular motion passes into molar motion
whenever the free separation of the moving molecules
is interfered with by the cohesion of enclosing masses.
Thus the steam in a cylinder pushes up the piston by
its expansion; the freed nitrogen in a discharge of

gunpowder in like manner pushes out the ball ; and the energetic movement of a heated gas bursts the vessel within which it is confined. Molecular motion also passes into atomic motion in decomposition by heat, and into electrical motion in the friction machine. Atomic motion passes into molecular motion when heat is generated by chemical combination. It also passes (apparently) into electrical motion in the galvanic current. Electrical motion passes into molecular motion when an interrupted current produces heat. Light, which is a phenomenon connected with the ethereal medium, must be neglected for the present.

This relation is quantitative—that is to say, a definite amount of Potential Energy passes always into a definite amount of Kinetic, and *vice versa*, while a definite quantity of each species is equivalent to a definite quantity of each other species, in either Mode. The law of conservation may therefore be subsumed under the following formula, where A stands for Potential and B for Kinetic Energy; 1, 2, 3, and 4 for the Molar, Molecular, Atomic, and Electrical species, and 5 for the Kinetic Energy of the ether (of which more hereafter) :

$$A1 + A2 + A3 + A4 + B1 + B2 + B3 + B4 + B5 = \text{a constant quantity.}$$

But while the total of Energy, like the total of Force, is thus constant, the total of each mode and species varies from moment to moment. Whereas the total of each Species of Force is as constant as the sum of their totals.

Again, while each unit of Force is rigidly bound up with each atom of matter (with which it is perhaps identical),[1] each unit of Energy may pass from one mass, molecule, atom, or electrical unit to another. It may also pass from matter to the ethereal medium, and *vice versa*. This can only happen, however, to Energy in the Kinetic Mode.

A mass in motion parts always with portions of its motion to all other bodies with which it comes in contact. It does so either by imparting to them a portion of its motion in the molar form (as when one billiard ball strikes another), or in the molecular form (as when heat is generated by friction). Hence every moving mass tends to part with all its Kinetic Energy more or less quickly, according as it is more or less impeded in its motion by more or less cohesion and gravitation. Thus a cannon ball parts with all its Molar Kinetic Energy at once when it strikes an iron target, and very quickly when it is fired in the air; a billiard ball parts with it more slowly, as it hits the other balls and the cushions; a quoit on ice

---

[1] It is possible to regard each atom as a centre of Force (i.e. Aggregative Power) liable to separation from other centres by means of Energies (i.e. Separative Powers).

more slowly still, as it meets the resistance of the air
and the gentle friction of the ice ; while a pendulum
under an air pump hardly parts with it perceptibly
by friction on its knife-edge, and a planet only by infinitesimal decrements to the ethereal medium.   A
molecule in motion parts similarly with a portion
of its motion to every other molecule with which it
comes in contact. When the two molecules, however,
possess equal motions, or, as we oftener say, are at
the same temperature, the amounts of gain and loss
neutralise one another.   But when the motions of the
Molecules differ, the more energetic parts with a
disproportionate amount of its motion to the less energetic, until the Energies of both are equal.   Hence it
happens that whenever the molecules of any mass
have a higher Kinetic Energy than that of surrounding bodies, the motion of its molecules is imparted to
the surrounding bodies till a state of equality is
reached.   As to Atomic and Electrical motions, we
know too little of their nature to speak with any confidence, but we see at least that they also tend to
pass away from the bodies with which they were as-
sociated, and to assume the forms of light and
heat.   In short, without fully anticipating the
chapter on the Dissipation of Energy, we may say that
whenever masses, molecules, atoms, or electrical
units are free to act in accordance with their aggre-

gative tendencies, without interference of antagonistic Forces or restraining power of continuous Kinetic Energies, they immediately unite, and impart their former Potential Energy in the Kinetic Mode directly to surrounding bodies, and ultimately to the ethereal medium.

We may thus summarise the contents of the present chapter : the sum total of all Energies in the Universe is a constant quantity ; and whenever one mode or species of Energy disappears it is replaced by an equivalent quantity of another mode or species.

CHAPTER X.

THE two generalisations briefly stated in the two preceding chapters under the titles of ' The Persistence of Force' and 'The Conservation of Energy' may be summed up under a still wider generalisation to which we shall apply the title of 'The Indestructibility of Power.' It may be formulated as follows.

The total amount of Power, aggregative or separative, in the Universe, is a constant quantity, and no Power can ever disappear or be destroyed.[1]

This short chapter cannot be enlarged by the addition of any further remarks. Like our first chapter on Power generally it does not admit of amplification.

---

[1] Readers of Mr. Herbert Spencer's *System of Synthetic Philosophy* will doubtless observe that it is this ultimate generalisation to which he refers under the style of *the Persistence of Force*, and not either of the minor generalisations subsumed under it. The author, however, makes this statement solely on his own responsibility, and has no warrant from Mr. Spencer for doing so. It is not improbable that Mr. Spencer would energetically dissent from acquiescence in the statement.

F

CHAPTER XI.

THE MUTUAL INTERFERENCE OF FORCES.

As the various portions of matter, molar, molecular, and atomic, all possess Forces of their own, it must necessarily happen that many bodies or particles are attracted in different directions with varying intensities by surrounding bodies or particles. Hence arises a certain cross-attraction or Mutual Interference of Forces. We shall consider in regular order the various modes in which each species of Force is opposed by interfering Forces.

Molar Force may be opposed to another Molar Force when two neighbouring masses each tend to attract a third mass. If all three masses be in every respect free—that is to say, if there be no other restraining Force, and no continuous Energy of relative motion—the three masses will aggregate simply. But in the large planetary bodies exposed to our observation the orbital Energy counteracts all the Forces; and we consequently see the sun, the earth, and the moon retaining their relative positions in spite of gravitation. There are certain instances, however, where the interference of Forces is seen, even in the case of Molar Forces. Thus, a large body like a table does not perceptibly attract even very small bodies on the floor, owing to the superior Power of the earth's attraction as a whole. Yet in the neighbourhood of much larger masses, such as mountains, a slight deflection of the plummet has been observed, because the attraction of the mountain has proved strong enough to counteract in part the attraction of the earth as a whole.

Molar Force is more commonly interfered with by Molecular Force or cohesion. A weight placed on a table or a ball suspended by a cord cannot aggregate with the earth generally, because the Force of gravitation is overpowered by that of cohesion. At a certain point, however, the Power of gravitation outweighs that of cohesion, and the table or the rope gives way.

We can scarcely say with any certainty that Molar Force is interfered with by Atomic and Electrical Forces : but there seems no reason to doubt that chemical attraction may act in opposition to gravitation by causing an atom to aggregate with another atom so as to raise it slightly above its previous level : while the position of a lump of iron suspended from a magnet (permanent or electro-magnetic) probably represents the interference of electrical

with molar Force. Our acquaintance with these phenomena, however, is so very superficial that it would
be premature to do more than hint at possible ana-
logies.

Molecular Force may be opposed by Molar Force
in the above-cited instances of a mass laid on a table
or hung by a cord. If the Molar Force overpowers
the Molecular, the table or cord breaks, and the mass
falls to the ground. One Molecular Force is opposed by another Molecular Force in the curious case
of what is called Molecular Tension. In such an instance, certain molecules on either side of a particular
set of molecules tend to draw it towards them,
and the stronger attraction finally succeeds in doing
so, leaving a disrupted portion on one side of the
line. Molecular Force is probably opposed by
Atomic and Electrical Forces, though here again no
very obvious instance can be cited.

Atomic Force is possibly opposed by Molar Force
as noted above. It is also possibly opposed by
Molecular Force; and this seems not improbable
when we recollect that many bodies will not combine
chemically unless at a high temperature—in other
words, unless their Molecular Force has been counteracted by an antagonistic energy. One Atomic
Force is certainly opposed by another Atomic Force
when two different atoms, each having affinity for a

third atom, are brought into close conjunction with it. This occurs in all ordinary reactions; and, as we see, the stronger affinity overpowers the weaker one. What may be the relations of Atomic to Electrical Force it would be premature even to guess.

Electrical Force as a whole is too little understood to permit of definite treatment. We may conjecture, however, that it is similarly affected with other Forces. In one case, at least, we can feel sure of an analogy. One Electrical Force can be opposed to another by placing two balls, pretty equally charged with Positive Electricity, opposite to one another, and at equal distances from a ball charged with Negative Electricity. In this case we set up a state of cross-tension like that of the interfering masses, the molecular tension, or the rival chemical affinities: and any slight difference in the two attractions will cause the one to outweigh the other. It would also seem as though, in the case of a Leyden jar, the molecular Force of the glass opposed the Electrical Force which tends to aggregate the opposite electricities: for when the Electrical Force reaches a very high pitch, the electricities escape from some point on the metal surface, and leave a hole pierced through the glass. The analogy of this case to that of the broken rope or table is obvious. On the whole, however, the subject is still too ill-correlated with other departments of physics to allow of positive statements.

In all the cases where the interference of Forces produces an actual separation between masses or particles previously in (relative) contact, it might at first sight seem as though there were really an exhibition of Energy and not of Force. As in the case of aggregative Energies, however, a little consideration will correct this idea. For the bodies always follow the *stronger* Force; and the result is, a total of closer and more intimate aggregation than that which before subsisted. If the cord can resist the power of gravitation, then the union between its molecules is a more intimate one than that which would result from the aggregation of the ball and the earth. If, on the other hand, the cord cannot resist it, then the total of aggregation is increased by the fall of the ball. So, too, if a body in chemical combination with another body can resist the affinity of a third body brought near it, the existing union is shown to be a closer one than that proposed for it. If, on the other hand, it cannot resist it, then the new union proves itself thereby to be closer and more intimate than the previous one. When we come to consider the material universe as an aggregating total, whose separative Energy is being imparted to the ethereal universe, this point will become much clearer.

CHAPTER XII.

THE SUPPRESSION OF ENERGIES.

WHEN a set of particles possessing Kinetic Energy is entirely surrounded by other particles, bound together by Force, it is possible up to a certain limit to suppress the Energy of the contained particles by limiting their mutual movements ; whereupon the Energy appears to exist in a dormant state. But when a certain point of suppression is reached, the Energy of the contained particles overpowers the Force of the containing particles, and a disruption takes place. Such a disruption is commonly known as an Explosion. Or again, at a point short of disruption, such an active separative impulse exists amongst the contained particles, that if any aperture be made in the containing wall, the contained particles rush out with Explosive Energy.

The abstract statement of this principle must be enforced by a few concrete examples.

The boiler of a steam-engine is a wall or partition

of molecules, rigidly bound together by cohesion. Within it, is a mass of water and steam, which is being raised to a high pitch of molecular motion by the fire underneath. Up to a certain point, it is possible to suppress or restrain the separative Energy of the steam by opposing to it the cohesive Force of the iron wall. But when a certain point of suppression is reached, the Energy outbalances the Force, and an Explosion takes place. At a point short of the Explosion, it is possible to open a valve and 'blow off steam': the energetic particles then rush forth with Explosive Energy. Similarly when a gas is reduced by pressure to the liquid state. Up to a certain point the Energy of the gas is suppressed; but when that point is passed, the Energy outbalances the Force, and an Explosion takes place. Short of the Explosion, it is possible to open the vessel, whereupon the gas rushes forth with Explosive Energy.

It is possible that certain (so-called) chemical combinations are really of this nature. Thus, certain compounds of nitrogen are very apt to explode, and it would seem not unreasonable to suppose that in their case the Energy of the free gas may be in some way confined by the combining atoms : while a match or other detonating agent may be the analogue of the valve or the stopcock in the above cited cases. This possibility will be more fully discussed in the succeeding chapter.

It is important to notice that one Energy may be opposed to another in producing a suppression. Thus Energy is expended in compressing a gas or bending a bow (a case which will be fully considered hereafter). So that just as Forces interfere with Forces, Energies sometimes oppose Energies. A suppressed Energy is regarded in the ordinary text-books as Potential. It is clear, however, that it cannot be so regarded from our present standpoint. It is essentially Kinetic, though its Kinesis is masked by surrounding bodies.

CHAPTER XIII.

LIBERATING ENERGIES.

WHEN any body or particle possessing Potential Energy is prevented from aggregating with any other body or particle which attracts it, by the interference of an antagonistic Force, its Energy can only assume the Kinetic Mode through the intervention of some external Energy. Such external Energy is itself necessarily in the Kinetic Mode. It is known as a Liberating Energy.

Put in more concrete language, this principle may be otherwise stated thus. A body can only be disengaged from the attraction of one Force and brought under the direct influence of another, by some movement affecting it. A moment's consideration will make it clear that this is a corollary from previously stated laws.

As we saw that the stronger Force necessarily outweighs the weaker, and as Forces cannot increase or decrease in intensity, the only manner in which any body or particle can be released from the Force which actually governs it and brought under the influence of another Force, is by some movement which either severs it from the sphere of the existing Forces, or brings it within the sphere of a stronger one. In the latter case, it is immaterial whether the movement brings the body into proximity with other bodies, or brings other bodies into proximity with it.

Molar Liberating Energies are those which release masses from the interference of a Force antagonistic to gravitation. The commonest instance of such a Liberating Energy is seen when we remove some obstacle which by its cohesion prevented the aggregation of gravitating masses. Thus a ball suspended by a thread is released by the separative Energy of a knife or scissors. A clock weight wound up but checked by a catch, is released through the Energy which removes the catch. A stone perched on a ledge is released by the puff of wind or the blow from a hand which causes it to topple over. A head of water confined by a sluice is released by the Energy which raises the sluice. A mass of ice on a mountain top is released by the Energy of heat, which breaks down the cohesion of its particles and allows it to trickle down the sides. Even in those cases where the intervention of the Energy is less apparent, we can see in an ultimate analysis that such Energy is really the moving

Power at work.   Thus, when the string decays instead
of being cut, it might seem at first sight that the cohesion melted away imperceptibly; but a closer consideration will show us that the dropping of water,
the action of heat and light, the approach of chemical
solvents in minute quantities, and the incidence of
other unobserved Energies is really the cause of the
decay.   So, too, if the water makes a way through
the sluice, or cuts a path for itself through the bank,
it can only do so by the slow action of incident
Energies, which wear away the cohering substance
that retains it.   And the stone can never topple over
from its ledge unless some animal pushes it, or some
slow water action wears away its supporting mass.
Molar Liberating Energies may also be seen in a few
cases where a chemical body undergoes a separation
which precipitates the heavier among its constituents.

Molecular Liberating Energies are those which
release molecules from the interference of a Force antagonistic to cohesion.   Two planed pieces of iron
cannot cohere if laid side by side on a table : they are
restrained in their places by gravitation.   But the
energy which apposes them to one another acts in
this case as a liberator.   In other instances, heat
performs the same function, by loosening cohering
molecules from their existing arrangement, and bringing them within the sphere of their mutual attractions,

as when we weld two pieces of iron by heating them, or by hammering them together. The contained energy of water fulfils a like office in gumming or glueing, and in mixing plastic clay or dough. In these cases, one cohesion has interfered with another, and the Liberating Energy, by causing a partial disengagement, finally permits the complete saturation of both affinities.

Atomic Liberating Energies are those which release atoms from the interference of a Force antagonistic to Chemical Affinity. Occasionally it is the mere Force of gravitation or cohesion which opposes this affinity, and in that case, the Energy employed in bringing the substances together is the liberating agent; as when we expose phosphorus to Chlorine. In other instances, however, the mere apposition of the elements is not sufficient, as when we expose carbon to oxygen; heat is then needed as a liberating agent; and we may conjecture that it acts by setting up such a molecular vibration in the carbon as takes each atom out of its existing stable arrangement with other like atoms, into a compound carbon molecule, and brings it within the sphere of the stronger affinity exerted by oxygen. This case leads on to those where the interference is between rival Chemical Affinities. The Energy which brings together two substances and permits the stronger affinity to overcome the weaker acts as a liberating agent. In this instance, too, heat is sometimes necessary as an additional factor, probably for the same reason as before. In the case of Cl and H light acts as the liberating energy. Other less obvious cases resemble those of a match, where friction performs the same function.

Electrical Liberating Energies are those which release Electrical Units from the interference of a Force antagonistic to Electrical Affinity. The usual vagueness of electrical science prevents any definite treatment of these phenomena; but we may consider the Energy which closes the circuit of a battery, or brings the discharging tongs to a Leyden jar, as essentially analogous to the cases noted above. Their fundamental similarity will be seen if we recollect that any Energy spent in overcoming the cohesion of the glass partition in the Leyden jar, and so causing it to break, would produce exactly the same effect.[1]

Under the head of Liberating Energies it will be

---

[1] The practised physicist will observe that a much wider significa-tion is here given to the term *Liberating Energy* than that which is usually attached to the expression 'Liberating Force' in the current phraseology of science. But if, as here contended, the cases are really analogous in every way, then there is no logical reason why they should not all be included under a single general name. Of course, if competent critics can point out any error in the conception here advocated, the classification falls to the ground; and throughout, it must be remembered that all the ideas contained in this treatise, though dogmatically stated for simplicity's sake, must be regarded merely as suggestions and points-of-view thrown out for the express purpose of placing the author's conception in a clear light.

convenient to include those other Energies which
act so as to permit the escape of suppressed Kinetic
Energies. Such will be the Energy which turns the
valve of a steam-engine or the stop-cock of a vessel
containing compressed gases. A more familiar instance is found in the Energy which draws the cork
of a champagne bottle. And if we were correct in
supposing an analogy between known suppressed
Energies and explosive nitrogenous compounds such
as gunpowder and nitro-glycerine, then the match or
blow which explodes them acts as an analogous
liberating agent. Liberating Energies of this description may be conveniently described as Liberators
of Suppressed Energies.

CHAPTER XIV.

MISCELLANEOUS ILLUSTRATIONS.

AFTER so long and so abstract an exposition, it may be well to give a few selected concrete illustrations, showing the interaction of the principles already laid down, before we proceed to those still more abstract and difficult problems which yet lie before us. We have hereafter to frame some clearer notion of the Relation between Ether and Matter, the Nature of Energy, and the Nature of Movement; which questions will require a power of abstract thought and concentration which is not possessed by every reader. But it may aid our comprehension of these highest abstractions if we more firmly grasp the concrete phenomena in which they are dimly manifested.

A lump of ice lies loosely on a mountain top. Its molecules are rigidly bound together by the Force of cohesion. The Force of gravitation tends to attract it, but the cohesion of intervening molecules interferes, and it cannot further aggregate, cannot get any lower, of its own accord. It possesses Potential Energy in virtue of its separation from the dead level of ocean : but that Energy cannot assume the Kinetic Mode so long as the interfering Force of cohesion prevents it. There are, however, various conceivable ways in which a Kinetic Energy may intervene to liberate it. The wind may blow it over ; a man may hit it with his stick ; or a peal of thunder may shake it down. In any of these cases, it will go down as a mass, all its molecules still locked together by cohesive Force. Again, the Kinetic Energy of ether (which we commonly know as Radiant Heat) may fall upon the mass, while still perched on its pinnacle. In that case, the motions of the ether particles will be communicated to the ice molecules, just as the motion of one billiard ball is communicated to another—or still more exactly, as one pendulum might set another in synchronous motion by striking it time after time. Under the influence of this separative Power, the molecules will slowly be unlocked from their cohesive union, and the ice will be melted. But the Energy which thus acts as separative to the molecules in their relation to cohesion, acts also as liberative in their relation to gravitation. The Potential Energy of each molecule (visible Energy of position) now becomes Kinetic, and they roll down the mountain

F

side in the form of water.   Let us suppose that
they unite on their course and make a cataract.
When they reach the level below (which for argument's sake we will suppose to be that of the sea) all
their Potential Energy has been transformed into
Kinetic.   Omitting the small amount lost by friction
on the way, this Kinetic Energy is immediately transformed once more, as the water reaches the surface,
from the Molar to the Molecular species.   It becomes
heat, and is radiated off into the surrounding
space.   Our ice has thus entirely parted with its
Potential Energy to neighbouring bodies, and to the
ethereal medium, though the water which represents
it still holds all the Kinetic Energy which originally
melted it.   It cannot again be raised to the mountain
top without the integration of fresh Energy. Whence is
this to come ?   In the majority of cases it is supplied
by the Radiant Heat of the sun.   This Energy, working upon the surface of the sea, causes separation
amongst its superficial molecules, which thereupon
rise into the air.   Thus we see that the same Energy
which overcomes the faint cohesion of the water also
overcomes in part the force of gravitation.   The
heated molecules, being less attracted than the colder,
are pushed upward by their pressure, and rise to a
considerable height.   The agent in raising them is
Energy.   So that the very same motion which keeps

the planets from aggregating with the sun, keeps the water molecules from aggregating with the earth. So long as they retain this Energy they continue to float at a great height. But they cannot retain it long. The surrounding objects at that height are very cold—in other words, are not in a state of high molecular vibration. Accordingly, when the molecules encounter a cold mountain top, towards which they are attracted by molar Forces, they part with their heat and aggregate under the influence of cohesion into ice. Their Kinetic Energy is now all gone, and nothing remains to them but the Potential Energy of their separation from the level of the sea. And then the whole cycle of changes begins over again.

Let us look next at a totally different instance, that of a cross-bow. This is a common illustration with physicists, and it has already been once hinted at, but no detailed explanation was given, because it will be presently seen that the case is much more complicated than at first sight appears. The Kinetic Energy of human muscles is employed in pulling the string back to the notch. The bow is then bent. Now this bending implies two forms in which the energy becomes dormant, which answer to the common expressions, *tension* and *pressure*. The molecules in the convex portion of the bow are pulled slightly apart from one another, but not beyond the sphere of their mutual attractions. We have consequently here a state of Potential Energy due to the separation of particles strongly influenced by cohesive Force. The molecules in the concave part of the bow, on the other hand, are pressed closely together upon one another by the Energy employed, which here acts in opposition to the Kinetic Energy of the molecules, whose natural vibrations are thus in part suppressed. Accordingly we have here a state of suppressed Energy. Both of these of course tend to become Kinetic, but are prevented by the interfering cohesion of the string and the trigger. The separative nature of the Energy employed is clear from the fact that if the string is pulled too far back, the strain upon the cohering particles becomes too great, and the sphere of their mutual attraction being transcended, they break apart with a snap. In the present case, however, having merely bent and bolted our bow, we have all our Energy bottled up in a dormant state. Next, let us release the string. The Energy which we employ in doing so, acts as a Liberating Energy with reference to the Potential Energy of the convex part, and as a Liberator of Suppressed Energy with reference to the concave part. It removes that cohesive obstacle, the trigger, which interfered with the mobilisation of the dormant Energies. The Molecular Force of cohesion now draws together the separated molecules of the convex part, and their Potential Energy becomes Kinetic. Through the medium of the string it is communicated to the arrow. The arrow flies rapidly through the air, parts with a small portion of its Energy by friction, but retains most of it till it pierces the target. Here, part of its Energy is used up in producing separation between the particles; while the remainder is given off in the form of heat. And so all our Energy is once more yielded up from its original possessor, the bow.

Again, let us take a case where chemical activity is concerned. A lump of coal possesses Potential Energy in the separation of its atoms from those oxygen atoms towards which it is attracted by Chemical Affinity. So long as they are merely in mechanical conjunction with one another, the interference of some other Force (probably cohesion) prevents them from aggregating. But when a Liberating Energy is applied in the shape of a match, the atoms rush together in a mutual embrace. Their Potential Energy becomes Kinetic, and they aggregate. But the Energy of their separation is not destroyed. It is communicated to the ether as Radiant Heat. In this state it may either pass away from our earth altogether, or it may be communicated to other bodies, in which case it is said to be *absorbed.* Let us suppose it is absorbed by a boiler of water. The water molecules

are then thrown into a state of vibration, which rapidly
severs them from one another until they assume the
form of steam. If this steam is allowed to issue from the
boiler, it will rapidly give off its Energy to neighbouring bodies, the ether included, and the Energy which
first passed from the coal and oxygen to the water,
will now pass from the water to the ether.   But
we may use the boiler to turn an engine.   In this
case part of the Molecular Kinetic Energy is transformed to the Molar species in the piston.   It is then
used up in initiating movements in the wheels and
cranks, all of which are finally retransformed into the
Molecular Species by friction.   If the engine is sta-
tionary, the friction will be between its parts ; if locomotive, between its parts and the rails.   Ultimately,
in every case, all the Energy is yielded up to the ether
in the Kinetic Mode and radiated off into space.

Now what is the conclusion which all these cases
force upon us? That whenever Forces succeed in aggregating masses, molecules, atoms, or electrical units,
the Energy of their separation, passing into the Kinetic
Mode, is transferred to surrounding bodies, and after
many or few vicissitudes is finally handed over to the
ethereal medium.   This is the point which must next
engage our attention.

CHAPTER XV.

THE DISSIPATION OF ENERGY.

IN the definition of Force given in our second chapter, a Force was stated to be a Power which initiated aggregative motion and resisted separative motion in two or more particles of ponderable matter, and possibly also of the ethereal medium. In the definition of Energy, given in our third chapter, an Energy was stated to be a Power which initiated separative motion and resisted aggregative motion in two or more particles of ponderable matter *or of the ethereal medium.* This addition in the latter case, and its qualified omission in the former, was intentional and significant. Though we cannot dogmatically say that the ether does not possess Forces, we do not know it to possess any; and if it does, the resistance which they offer to separation appears to be almost infinitesimal. It may well be that ether is merely a more tenuous kind of matter, animated by the same Power as the ponderable bodies : but even if it is, we know that it can be conspicuously affected by Energy, while we do not know that it can be conspicuously affected by Force.[1] From this difference flows a very important corollary which may be formulated as follows.

The Energy liberated from the Potential Mode when bodies or particles aggregate under the influence of Force tends ever to assume the Kinetic Mode, and to be transferred from ponderable matter to the ethereal medium.

As Liberating Energies are perpetually setting free Potential Energy, and permitting aggregative motions to be set up under the influence of Force, and as the Kinetic Energy thus liberated is transferred to adjacent bodies, a part of it at least must be transferred to the ether. Furthermore, as that part of it which is transferred to the ether is radiated off in every direction into space, it must happen that the greater part of it is lost for ever to ponderable matter, and imparted to the ethereal medium. For, although some portion of the Energy may meet in its course with ponderable matter, and be absorbed thereby, yet inasmuch as the interstellar spaces are vastly larger than the interspersed ponderable heavenly bodies (so that in most

---

[1] The mere fact that motion can be lost by ethereal friction, as we know in the case of heated molecules and believe in that of planetary bodies, would lead us to suppose that the ether has something resembling cohesive Force.

directions motion may be continued in a straight line for ever without meeting one) it necessarily happens that the greater portion will never meet with any ponderable matter, but will go on, presumably *ad infinitum*, traversing the ethereal medium. This principle, which will be fully expounded in its concrete aspect in Part II. of this work, must at present be accepted in this its abstract aspect, on the ground here laid down.

Again, though much radiant Energy may fall upon any one mass from another (as on the earth from the sun), yet inasmuch as this Energy is the correlative of an aggregation which has taken place in the radiating mass, whereby some of its Potential Energy has become Kinetic and been imparted to the ether, and inasmuch as the portion absorbed by that particular mass bears an infinitesimal ratio to the portion radiated *ad infinitum*, it must follow that on the whole every aggregation involves a loss of Energy to ponderable matter and a gain of Energy to the ethereal medium.

Once more, even that portion of Energy which is absorbed by any other mass will in part be used up in Liberating Energies (as when solar heat melts a piece of ice on a mountain top), and will accordingly itself be a cause for future transfers of Energy from ponderable matter to the ethereal medium. And finally, this absorbed Energy itself will in part at least be radiated off from the absorbing mass and imparted once more to the ethereal medium.

We thus see in every case that all Energy tends to be lost by ponderable matter and transferred to the ethereal medium.

CHAPTER XVI.

THE NATURE OF ENERGY.

WE now come to the most abstract and fundamental question of all. What is the true nature of Energy ? In the beginning of this book we took it for granted that Force was equivalent to Aggregative Power and Energy to Separative Power. That first assumption, however, is in reality the point which our treatise is meant to prove, and we have tried to prove it by running through the chief manifestations of Power and showing how simply and truthfully they can all be formulated on this principle. We have endeavoured, in other words, to point out the perfect congruity of our assumption with fact. Having done so first in the abstract, we shall proceed to show how the phenomena which form the subject-matter of the various sciences, and how the concrete course of events in the universe, can be expressed in terms of our formulæ. But before we go on to these departments of our subject, we must try to gain a clearer conception of the real nature of Energy.

Energy is Separative Power. Every Energy in the Universe was primordially a mere statical separation of masses, molecules, atoms, or electrical units. If there were no such power as Force, every one of these bodies would have remained for ever apart and immovable. But as forces draw together these mutually attractive material objects, their Energy assumes for a moment the Kinetic Mode. The falling water, the moving glacier, the oxygen rushing to unite with the coal, each pass for a shorter or longer period through the Kinetic stage. As they aggregate, their Energy is given off in some other form of motion, involving separation. But as they move about, they part with this motion in separating other masses or molecules, and the attractive Force soon brings them together again.

And what is the meaning of the Law of Conservation? Simply this : that the total of statical separation, plus the total of motion, in all particles whatsoever, material or ethereal, is always a constant quantity. In other words, separation can never yield to aggregation without. generating an equivalent amount of motion, which itself may pass into separation of some other sort : while, conversely, motion can never cease without generating either an equivalent separation or an equivalent other motion. Thus a body at a height cannot fall without generating an amount of motion proportionate in a known ratio to its mass and height, which motion itself is transferred on the body's fall to its several molecules, causing a separative action among them; and this motion is again transferred to the ether: while, similarly, a piece of coal cannot combine with oxygen without yielding up its separation in the form of molecular motion, which motion may raise vapour of water to a considerable separation from the earth's central mass, and be itself finally yielded up to the ether.   In short, the alternative Modes of Energy are Actual Separation, and Motion which eventuates in Separation.

Furthermore, no body can be prevented from saturating its aggregative tendencies except by an Energy. There is no known way in which the total of bodies can be kept apart from their closest conjunction with one another except by continuous motion, like that of a planet, a top, or a vibrating molecule. Even the case of Interference of Forces, which at first sight seems an exception, is not really so (though for convenience' sake we have treated it as such), because to suppose that the suspended ball will break its string or the weight push through the table is to suppose that a weaker aggregative tendency will overpower a stronger one. In all the larger bodies of the Universe we see the discrete state maintained by orbital Energy : and in all the molecules of liquids and gases on this earth we see the discrete state maintained by heat, or continuous vibration.

CHAPTER XVII.

THE NATURE OF MOTION.

LAST of all comes the question,—What is Motion ? Divesting our minds of all concrete associations and looking at the phenomenon in itself, we arrive at the following unfamiliar conclusion.

Motion is the Mode by which Energy (or Separation) is transferred from one portion of matter to another, and ultimately from matter to the ethereal medium.

When the Motion is simply separative we see this in a moment. A ball fired upward, a weight carried to a height, or an atom disengaged from a compound, show us motion as equivalent to separation, in its naked form. When we look at Motion along a line at equal distances from the attractive centres— as in the case of a locomotive running along a level —we do not at first see how the Energy can be considered as separative. But as soon as we reflect that the Energy required for the purpose is entirely relative to the resistances which must be overcome—as soon as we recollect that if there were *no* friction, the initial Energy would carry on the moving body for ever, and that where there is little friction the moving body continues to proceed for a long period in the same direction without conspicuous loss of speed—we see that each new increment of energy from the burning coal is used up—not in intensifying the rate of motion, but in overcoming friction, in wearing down the projecting particles of the machinery and the rails, in producing heat, and so, ultimately, in setting up separative actions.   This case leads us on to that of a planet having orbital Energy, or a molecule having Vibratory Motion.   In both these instances the substance to which the Energy is imparted is far subtler and more tenuous, being in fact the ethereal medium.   Yet in both we see that as their Energy is lost, they aggregate with attractive bodies, and we thus perceive the separative nature of their motion.   At the same time we see it as a mere incident in the transference of separation from matter to ether.   Lastly, in the case of aggregative movements, we see that the Motion replaces for a time the separation of masses, molecules, atoms, or electrical units, as they rush together; but we also see the same separation afterwards transmitted to some other form of matter or to the ether, as heat, light, elec-

trical separation, or some other form of separative
Energy.

Again, in every case, the ether is the final gainer
of Energy, and every Motion is only an incident which
ultimately effects the transfer of Energy (i.e. separation)
from matter to ether.   On the surface of our earth,
where so large an amount of Energy is being daily
poured down by the sun, this truth is masked by the
fact that new Energy continually replaces the old.
But if we leave out of consideration the accretions
thus made to our store of Energy, we shall see that
every Motion originates in an aggregation—whether
it be through the fall of a body at a height, or the
burning of coal in an engine, or the oxidation of
food in an animal body—and that after the motion
has taken place, there is a less total of Potential Energy
on the earth, while the Kinetic Energy has been transferred, in whole or in part, to the ether.   This principle, here briefly alluded to in the abstract, will be
fully developed in the portion of this work devoted
to concrete phenomena.   Far more evident, however, is this truth when we look to the wider sidereal
system.   There, we see at once that all Kinetic Energy
is the correlative of an aggregation, and that the
separative Power, which once divided the ponderable
matter composing the various suns, is now being
radiated off, as they aggregate, in the form of ethereal
Kinetic Energy — or, as we oftener say, of Light and

Heat. This Energy, when it falls upon such a mass as our own planet, at once displays its separative nature by such phenomena as the melting of ice, the raising of aqueous vapour, the formation of winds, and the production of living organisms. These questions, again, will be fully discussed in the Second Part of this book.

Briefly, we may say that the shortest formula to embrace the facts of Kinetic Energy is the following:—Motion is the redistribution of separations.

We have now completed our rapid survey of the abstract principles of Transcendental Dynamics, and may proceed to consider their concrete manifestations. Before doing so, it was the author's original intention to glance briefly in a separate Part at certain other subordinate facts connected with the development of the subject. The Laws laid down in the present First Part mostly refer to that department of science known as Physics; though we have treated incidentally of many facts commonly looked upon as chemical and electrical. A special Part ought to have been dedicated to a brief examination of certain qualitative propositions in Chemistry and Electrical Science: but this task, unfortunately, the author has found impossible of achievement with his existing knowledge. He therefore proceeds at once to the concrete manifestations.

Part II

CONCRETE OR SYNTHETIC

CHAPTER I.

WE have now to consider in their concrete applications the abstract laws laid down in the First Part. Our chief object in doing so will be to show how simply and clearly the wider dynamical relations of the Universe can be comprehended under our conception of Force and Energy, as two mutually opposing and indestructible forms of Power.

If we conceive a Universe without any inherent Force, all of whose atoms stood at varying distances from one another, we can see that it would continue for ever motionless, all its Energy remaining in the Potential Mode as simple statical separation.[1]

---

[1] In the current language of Physics, such a state of separation would not be spoken of as Potential Energy at all. It would only be considered as such when a Force attracting the atoms had been introduced into the closed system. It is unfortunate that we must use the term 'Potential' in such a case; but as we have here kept it throughout, instead of the simpler and more logical phrase 'Energy of Statical Separation' proposed in an earlier page of this treatise, it will be better, in spite of the verbal incongruity, still to continue its use in this Part

There would be nothing to draw together its scattered parts, or to set up motion in a single one of its particles.   On the other hand, if we conceive a Universe actuated only by Force, we can see that it would aggregate immediately if it were in a discrete form, or that it would preserve its inertia if it were already absolutely aggregated.   There would be no Conservation of Energy, and each mass, as it glided into the central whole, would simply subside without communicating its motion or separation to adjoining masses.   But the only Universe which we know by observation is actuated both by Force and Energy.   It consists in part of ponderable atoms, molecules, and masses, each of which tends to aggregate with the others, but each of which can only get rid of its separation by passing it on to some other substance, either as separation or as its equivalent, motion.   It also consists in part of other relatively imponderable particles, known as ether, occupying all the interspaces, great or small, between the ponderable bodies, and capable of receiving and imparting Energy from or to the ponderable units.   And inasmuch as all moving bodies part with some portion of their motion to every other body with which they come in contact

Of course the word 'Energy' itself ill describes such a Power as that which we have envisaged under that name—a power whose chief primordial manifestation is wholly statical. But we have thought it well to continue calling it by the most familiar name.

in every direction, and, further, inasmuch as the interspaces of ponderable bodies are infinitely greater than the space occupied by such bodies, it must necessarily follow that the total amount of energy received by the ether from all ponderable bodies must be very much greater than the total amount of Energy received by all ponderable bodies from the ether. In other words, the ponderable bodies must be aggregating, and passing on their Energy to the ether.

Our Dynamical Formula of the existing Universe, so far as it is known to us in its present stage, will therefore be a deduction from the Law of the Indestructibility of Power—that is, from the joint principles of Persistence of Force and Conservation of Energy. It may be stated as follows.

All the ponderable bodies of the Universe are continuously aggregating under the influence of Forces, and are imparting their Energy to the ethereal medium: such continuous aggregation being only locally interfered with where the Energy imparted to the ether by one aggregating mass falls upon the surface of another mass, and there sets up temporary separative action, in opposition to the local Forces.

It may be added that such local separative action, as hinted above, is not sufficient in amount to counter-

act the general aggregative action, and that, in consequence, the ponderable matter of the Universe is
daily becoming, as a whole, more aggregated, while
the ethereal medium is daily becoming more energetic:
though we have no means of knowing whether the
Energy absorbed by the ether continues always in the
Kinetic Mode, or finally assumes the form of statical
separation.

We have now to apply this Formula to the explanation of the concrete phenomena presented by the
Universe in its various portions. Our explanation
will serve a double purpose, as a deductive affiliation
of the several sciences on the Law of the Indestructibility of Power, and as a verification of our abstract
principles by their exact coincidence with well-known
facts.

CHAPTER II.

THE SIDEREAL SYSTEM.

THE life-history of the material Universe, as known and predictable by us, is the history of its passage from a diffused nebulous state to an aggregated solid state, through a vast number of intermediate stages, each one of which is an advance in aggregation upon the stage which preceded it. Whether or not the universe had any previous state to that of the earliest known nebula, whether it will have any later state than that of the absolutely aggregated mass, are speculative questions into which we will not enter in the present treatise. It will be sufficient for our purpose to trace the history of matter in its existing phase, from its first known form as numberless diffused atoms, to its last knowable form as a single aggregated mass.

All modern science compels us to posit as starting point a primordial state of the Universe in which its various masses, molecules, and atoms stood apart from one another at unknown distances. But each particle had inherent in it those forces which were destined in the future to effect its aggregation with every other. Accordingly, however we figure to ourselves the beginning as absolute or relative (and it must be allowed that we have here reached the utmost limits of our conceptive faculty), we cannot but suppose that from the moment of that beginning the various particles began to set at once towards one another. The primordial form of Energy was therefore all Potential, and it must immediately have begun to assume in part the Kinetic Mode.

If we assume that the primitive cosmical nebula was perfectly spherical in shape, and absolutely homogeneous and uniform in the disposition of its various atoms, then we can hardly resist the inference that, as each particle would be quite evenly attracted towards the common centre of gravity, there would have resulted a single aggregating sphere, whose Potential Energy would all have passed into the form of heat as the atoms clashed together, and would have been slowly communicated to the circumambient ether, until the whole mass had assumed its most aggregated shape. But as we do not find this condition fulfilled by the existing Universe, we must conjecture that the primitive nebula was not uniform in shape or in texture—that it enclosed within it groups

of tenuous matter more or less separated from other
groups by lines of demarcation more or less strong.
Such inequalities of distribution may have been infinitesimal; for it is only necessary to our purpose
that certain atoms, besides their general gravitation
towards the common centre, should also have dis-
played a special gravitation towards special centres.
Granted this, the reason for the discrete condition of
the sidereal masses becomes obvious.

But when each ultimate particle began to unite
with each other particle, the Law of Conservation de-
manded that their Energy of statical separation should
not die out as they coalesced, but should pass on to
some other body or assume some other form. The
manner in which it actually shows itself is that of
heat. Within each sidereal mass, the atoms are in
a fierce state of vibratory movement, the correlative
of their previous separation and of the Kinetic Energy
of their mutually aggregative motion. This vibratory
movement is perpetually being communicated to the
adjacent ether as Radiant Energy, and a corresponding aggregation within the sidereal mass is perpetually
taking place. Each sun is thus a body of aggregating atoms, being drawn together from moment to
moment by their inherent Forces, while their Energy
of statical separation, after passing into the continuous
Kinetic form as true Heat (molecular vibration), is

yielded up, little by little, to the adjacent particles of
ether as Radiant Energy. The Energy thus absorbed
by the ether is passed on by it from particle to
particle of its substance, radiating in every direction
throughout all space. Some small portion strikes
the surface of our planet, both from our own sun
and others like it; and it is known to us as Light
and Heat.

We thus see that all the Energy of Light and
Heat radiating through the whole of space from the
various material masses has for its origin the statical
separation of the primordial nebula. But is this
equally true of the Kinetic Energy of their various
relative motions? There is reason to think that
it is.

The Universe as a whole has a common centre of
gravity, towards which all its various masses are
attracted. Those masses still possess Potential
Energy in virtue of their separation from one another
and from this central point of union: and it is clear
that if they were to aggregate suddenly round that
point, their Potential Energy would become Kinetic
as they fell, and would be transmuted into Heat as
they clashed together at the common cosmical meet-
ing-place. It would then be radiated off into the
ether, and the matter would gradually assume a solid
and perfectly aggregated form. Now, it is possible

that some of the sidereal masses may be thus gravitating towards the common centre in a direct line ;
and if they are, then it is clear that their motion is
the correlative of their previous separation.   But
it is more probable that the various suns are prevented from aggregating directly with one another
by some form of continuous motion.   We are sure
in the case of the best-known large masses—the earth
and other planets—that they are prevented from
aggregating with their relative centre, the sun, by
the continuous Energy of their orbital motion.   We
also know that certain special suns—the double stars
—have such a relative motion with regard to one
another.   We further know that all stars have a
proper motion whose cycle is so immense that it
cannot be measured by the short period of human
observation.   It is probable, therefore, that the
ascertained cause which prevents central aggregation
in the known cases (namely, orbital motion) may be
fairly extended to the unknown cases.   We may
conclude, accordingly, that all the heavenly bodies
are prevented from aggregating around the common
cosmical centre of gravity owing to their possession
of a relative orbital movement.   Of course, there may
be many cycles of such orbital movements one within the other, as we know to be the case with the
satellites which circle round a planet, while the plane

circles round the sun, and the sun has his own proper motion. All that is contended here is merely this —that each mass or set of masses is probably prevented from aggregating with each other mass or set of masses, around their relative centre, or around the absolute cosmical centre, by some continuous Kinetic Energy, analogous to the known orbital motion of the planets and their satellites. Now, is this continuous Energy the transmuted form of previous separation in the parts of each mass? In the best-known case—that of the masses composing the solar system—astronomical authority has decided that it is. Laplace has shown that the orbital motions of the planets and satellites, as well as the axial motions of the sun and its dependent bodies, may be accounted for by the falling together of nebulous matter, whose Energy of separation, becoming Kinetic in the act of aggregation, and then receiving a check, communicates motion to the whole mass. In what way this axial motion is converted into orbital motion we shall see when we come to consider the solar system in the next chapter. For the present it must suffice to point out that the Energy of relative motion in heavenly bodies is thus probably due, like their Energy of Heat, to the primordial Potential Energy of their originally separate and discrete particles.

Again, is this Molar Kinetic Energy of relative motion in the various heavenly bodies being yielded up to the ether, as we saw to be the case with their Molecular Kinetic Energy, which is being dispersed from moment to moment through all space in the radiant form? There are reasons for thinking that this also is the fact. It is now pretty generally admitted that orbital Energy is slowly lost by ethereal friction in the case of the planets: and there is no reason to doubt that it is equally lost in the case of these larger masses, the fixed stars. And as the Kinetic Energy of orbital motion seems to be the only barrier to aggregation under the influence of gravitation, it will follow that as this Energy is imparted to the ether, the various heavenly bodies will draw closer and closer together, until, when all their Energy has been transferred to ether, they will aggregate in absolute contact around their common centre.

Let us restate then, in a simpler way, the probable conclusions to which we are led. The Material Universe originally existed as a vast nebula of discrete particles, in which Persistent Forces and Conservative Energies were inherent. As its Forces drew together the particles into several masses their Potential Energy became Kinetic. Part of it assumed the Molar form, and drove the various masses in orbit

within orbit around their relative centres, and, ultimately, round the common cosmical centre. Part of
it assumed the Molecular form, and kept the molecules of each mass in a state of rapid continuous
vibration. But each Kinetic Energy alike was and is
continually being yielded up to the ethereal medium.
As Radiant Energy, the Molecular motion is from
day to day imparted to the ether, and transmitted to
the furthest ends of space.   Some small portion of it
falls upon other material masses, scattered like lonely
islands in the ocean of ether, and may there set up
separative action : but the vastly greater portion is
never again expended on a particle of matter.   As
ethereal friction, the Molar motion is more slowly
yielded up to the ether, in which it produces (probably) waves of heat (or low-powered radiant Energy).
And there is no reason to doubt that this process will
go on indefinitely until it reaches its final result. The
Molecular Motion will probably be dissipated until
each mass has grown cold and inert : the Molar Motion will probably be dissipated until all the masses
aggregate round their common centre.   The Material
Universe, which began as a number of discrete particles, will end as a single aggregated mass : all the
Energy which was inherent in its separate form will
have been transferred to the ether : and motion will
have been the means of transference, the repre-

sentative of the separation during its intermediate stage.

Of course, in this brief sketch of the cosmical life-history many incidents of vast relative importance are necessarily omitted. One mass—whether sun, planet, or satellite,—circling round another, may part with its Molar or orbital Energy, and may aggregate with its central mass, long before other masses have done so. At the moment when two such bodies aggregate, doubtless some portion of their Molar Energy will still remain, and this will probably be converted into the Molecular species, and radiated away as heat and light. But such minor incidents, forming the several steps of the great process by which matter is aggregated and Energy dissipated into ether, do not interfere with the main process as laid down above. Moreover, as the history of one such episode—that of the solar system—will be more fully considered in our next chapter, it is less necessary to enter into details at the present stage.

This chapter contains much that is purely speculative and may raise objections in many minds. That is inevitable, considering the vastness of the subject and our ignorance of the facts. But as we progress to the solar system the speculative character of our explanations will gradually decrease : and when we reach our own planet—the most practically important

H

to ourselves—the element of hypothesis will disappear altogether. For symmetry's sake, however, it is necessary that the less certain cosmical application of our principles should precede the more certain mundane application.

CHAPTER III.

THE SOLAR SYSTEM.

AMONG the minor aggregating masses into which the cosmical nebula may be supposed to have split up, in the course of its general aggregative cycle, was a group of matter out of which our own solar system has been developed. In its earliest separate phase we may suppose this group to have consisted of various minor portions, in varying stages of aggregation, revolving in a single direction around their common centre. (Details about the direction of Neptune and Uranus may be safely neglected.) We may further suppose that the nebulous or quasi-nebulous mass thus composed again divided itself, along its weakest cohesive lines, into other portions, which have aggregated into the existing planetary groups; while these latter again subdivided themselves into the central masses and their satellites. It is immaterial for our purpose whether, with the earlier evolutionists, we regard these changes as taking place in a relatively homogeneous substance, a diffused nebula, or whether, with their later followers, we set them down to aggregative action in comparatively solid and discrete masses (meteors), like those which we know to exist in large tracts within the sphere of the solar system.   But the important point to notice in either case is this, that these groupings and sub-groupings took place under the influence of Forces, and that the Potential Energy of separation between the masses or molecules became Kinetic as they clashed together, and assumed the form of Heat.   The various masses thus became each of them a little sun, aggregating around their several centres, and radiating their Energy into the surrounding ether.   As in other cases, some small portion of this Energy would fall upon neighbouring masses, belonging either to the same system or to other systems, and would there do a little towards retarding the aggregation of their matter and the dissipation of their Energy ; but the greater portion would doubtless be lost in the vast interstellar spaces ; so that the general result would be a loss of Energy to matter, and a gain of Energy to the ethereal medium.

The various planets and satellites thus formed would still possess Potential Energy in virtue of their continued separation from one another.   They would also possess Molar Kinetic Energy in virtue of their orbital movement, which acts as a preventive to their immediate aggregation with their common centre, the sun. And, finally, they would possess Molecular Kinetic Energy through the vibratory movement of their molecules, derived from the previous Kinetic Energy of their aggregative motion. But as their particles vibrated, they would part from moment to moment with portions of their Energy to the surrounding ether. This loss of Energy would only largely affect the periphery of each mass, and would at first be inconspicuous at the centre. It would also affect the smaller masses much more rapidly than the greater, for two reasons; both because the amount of aggregating matter being less, the amount of heat generated would also be less; and because losses of heat from the periphery could not so easily be made up by conduction from the centre. The smaller masses would accordingly cool at their surfaces at comparatively early periods; while the larger ones, in proportion to the amount of unaggregated matter within the sphere of their attraction, would continue for long periods to receive fresh accessions to their molecular Energy, and to radiate Light and Heat into the surrounding ether. Especially would the largest mass of all, the central sun, continue for an immense era to aggregate the surrounding masses and to radiate the liberated Energy into the space around.

Further, we may conclude that as the surface of each mass parted with its Energy, its superficial molecules would be drawn together by the Forces of cohesion and chemical affinity. We should thus get a solid cohering framework on the exterior of each mass, as soon as it had parted with a considerable portion of its molecular Energy to the surrounding ether.[1]   Through this cohering crust, the central Energy would slowly escape by conduction : until, sooner or later, we might expect each such mass to consist of a cold and inert collection of molecules, the whole Energy of whose previous separation had been yielded up to the ether.   But a special incident of this transference might occasionally occur to break the monotony of its simple course. As the central Energy escaped through the cohering crust, there might be a tendency for the interior molecules to be drawn together under the influence of cohesion and gravitation. A second crust would thus tend to form itself beneath the outer one. Thereupon, the Force of gravitation might outweigh the cohesion of the outer crust, which would yield under the strain and fall in upon the subjacent mass, breaking along its line of least cohesion. Each such fall would be

[1] The cases of the ocean and the atmosphere, which appear to contradict this general statement, but whose form is really due to the separative action of radiant solar Energy, will be treated in the next chapter.

itself a change of Potential into aggregative molar
Kinetic Energy, as the masses fell together; and
when they clashed, the Energy would assume the
molecular form and be given off as heat. But, however the aggregation takes place, as the matter of
each group aggregated more and more closely round
its centre, the Energy of its previous separation would
be given off as heat, and would finally be imparted, as
in every other case, to the ethereal medium.

While each mass was thus parting (by radiation)
with the Molecular Kinetic Energy resulting from its
previous separation and aggregative motion, it would
also be parting, though more slowly (by ethereal
friction) with the Molar Kinetic Energy of its orbital
motion. Each satellite would thus be drawing pro-
gressively nearer to its primary, and each planet to
the sun. As every unit of Energy was lost, gravitation would draw the body one unit nearer to its
relative centre. It might thus be expected that each
satellite would aggregate with its primary before the
primary planet aggregated with the sun. At each
such aggregation, though the greater part of the
orbital Energy would doubtless be already dissipated,
yet it is probable that as the two bodies glided together (for they would not *fall*, as is often erroneously
said) there would be a considerable residue of Energy
still remaining, which, on the shock of collision,

would be converted into molecular motion (or heat),
and would be sufficient to reduce the bodies to a
molten state. But this incident, instead of interfering with the final aggregative process, would really
hasten it: because the new united body would at
once begin radiating off its heat into space, and once
more cool down to a mass of cold and inert molecules.
In other words, all the remaining Energy of separation belonging to the satellite in virtue of its discrete
condition, and all the remaining Kinetic Energy of
its orbital motion, would thereupon be dissipated into
the surrounding ether: and the united mass would
continue to gravitate slowly towards the central
sun. What is thus probable of the satellites in
relation to their primaries is equally probable of the
planets in relation to the sun. As their Energy of
orbital motion is dissipated by ethereal friction, we
conclude that they are drawing nearer and nearer,
step by step, to the centre of our system. And there
is no reason to doubt that they will continue to do so
until they each slowly aggregate with the central
mass, converting their remaining Energy as they
clash together, into heat, which will itself go for a
time to swell the volume of solar Energy, and will be
radiated off like the rest into surrounding space.
Finally, when the sun has aggregated with himself
all the matter of the solar system, we may conclude

that he will ultimately radiate off all the molecular Energy derived from their aggregation, and become himself a cold and inert mass, like some of those burnt-out stars revealed to us by astronomy. And then we may imagine that this single sphere, which contains all the matter of our system, will itself proceed to aggregate with some other mass, in that general cosmical group of which it forms an unimportant member. Of course, it is impossible to conjecture which of these aggregations will take place first; and it is quite conceivable that the whole solar system might clash with some other group of worlds before its various members had time to aggregate in this regular order with one another; but if our suggested theory of a general subordination of systems and cycles to a common cosmical centre be correct, then just as each satellite would aggregate with its primary before that primary had time to aggregate with the sun, so each planet would have aggregated with the sun before the sun had time to aggregate with its superior mass. However this may be, it will be sufficient if we regard the probable course of events in the solar system as a specimen of the probable incidents everywhere accompanying the course of aggregation round the common cosmical centre, and briefly hinted in the preceding chapter.

At the present moment of time, we occupy a
middle point in the systemic epoch thus sketched out.
The sun, our central mass, is still in a state of rapid
molecular motion, which he imparts as Radiant
Energy to the ether. He has yet much outlying
matter to aggregate, and it seems probable that small
aggregations are from day to day taking place. Of
the planets, the smaller have cooled down sufficiently
to possess a firm and coherent outer crust, while the
larger are still in a very volcanic state. The satellites
have probably radiated away all their proper heat,
and are already cold and inert to their centres. The
surface of the most easily observed, our own moon,
shows the vast corrugations which result from the
continual collapses of the crust upon the once heated
nucleus, and the reactions of the molten interior upon
the coherent outer shell :—corrugations relatively (if
not absolutely?) much greater than any at present
found upon the surface of our own earth. Some
small fraction of the Energy radiated from the sun
falls upon the cooled exteriors both of planets and
satellites. Of this, the greater portion is reflected
back by dispersion, as we see from the case of the
moon, in every direction (only a small fraction of
this fraction again falling upon other masses). But
a certain lesser portion is used up in heating the outer
crusts, in setting up evaporation, currents, and winds.

and in producing the phenomena of organic life. These secondary separative reactions of radiated Energy upon the surface of a planet form the great mass of those phenomena which are generally observed by human beings.

CHAPTER IV.

THE EARTH.

As we pass from the solar system to our own planet, we leave the region of hypothesis, and arrive at that of known facts.

The earth is a collection of material particles, the vast majority of which, so far as revealed to our observation, are in a state of stable aggregation with one another, molar, molecular, chemical, and electrical. Its centre may be—and probably is—still occupied by a molten (though rigid) mass, whose heat has not yet been fully conducted away : but the outer crust, except at its exposed surface, consists of matter aggregated in those molecularly cohering and chemically passive masses known as rocks. Its exterior is not absolutely regular, but is in places corrugated into certain elevations and depressions which we call mountains, tablelands, valleys, and ocean-beds. The portions elevated above the general level possess Potential Energy in virtue of their elevation : but the

Force of gravitation being interfered with by that of cohesion, this Energy cannot assume the Kinetic Mode without the intervention of an external Liberating Energy. In short, while the centre of the earth may still possess some molecular Energy of its own, which is only slowly escaping through the outer crust, its hard exterior is for the most part thoroughly aggregated and almost devoid of relative Kinetic Energy, except so far as it obtains small daily increments from the solar radiation.

If for a moment we leave out of consideration the solar Energy thus absorbed, we can form some conception of the appearance which the earth would possess, supposing it left to its own resources. The whole ocean and all the other water on the earth would be frozen into a solid mass. There would be no plants or animals on the surface, nor any winds, storms, rain, snow, or lightning. What might be the condition of the atmosphere we cannot say ; but we may guess that it would be greatly diminished in size, if not absolutely reduced to a solid form. Motion upon the surface would be all but unknown : the only movements which could ever occur being those which would occasionally result from those internal causes that give rise to earthquakes and volcanic eruptions. These would still take place, as the gradual loss of Energy from the central mass would

make the Force of gravitation outweigh that of cohesion; and the Potential Energy which thereupon
would be mobilised might act as a liberative agent to
certain masses on the slopes, besides causing perhaps
a temporary melting of some small portion of the
solidified water through volcanic heat. But these
incidents would themselves only accelerate the loss
of the remaining proper Energy of our planet, which
would soon be imparted to the ethereal medium, and
leave our earth at last a perfectly inert mass of
aggregated particles.

In the world as we know it, however, very different phenomena are observable; and all these are
due to the separative action of Energies radiated from
the sun, which fall upon our earth, acting partly as
separative agents for the superficial molecules, and
partly as liberative agents for the various Potential
Energies whose mobilisation is prevented by interfering Forces.   Falling upon the atmosphere, the
Kinetic Energy of ethereal undulation prevents its
aggregation and keeps it permanently in the gaseous
form.   If it be objected that the non-absorption of
radiant heat by the gases of the atmosphere is opposed
to this view, it may be answered that actual absorption is not necessarily implied: it will be sufficient
for our purpose if the original molecular mobility of
the gases is not diminished by communication with

the ether. We cannot experiment upon the conduct of oxygen or nitrogen at the absolute zero of temperature, but we have no reason to doubt that at some extremely low point they would follow the example of all other bodies, part with their molecular Energy to the surrounding ether, and pass through the liquid into the solid form.[1] We know already that carbonic anhydride can assume the frozen condition, and it is hardly probable that the simple atmospheric gases would not do the same, under similar circumstances, could we only command sufficient Power for their liquefaction. Falling upon the water, the ethereal Energy acts in opposition to its cohesive Force, and keeps it habitually in the liquid state, at least in tropical and temperate climates. Nor is it only by compelling them to assume the gaseous and liquid forms that the ethereal Energy displays its separative nature on air and water: it also acts in opposition to gravitation. It heats many water-molecules till they evaporate, and then raises them to considerable heights in the air. It expands the atmosphere of the tropics (by conduction and convection), and causes the phenomena of monsoons, winds, and storms. In a similar way it produces the ocean currents. And it thus becomes

---

[1] Since this was written, the solidification of oxygen has been actually accomplished.

the cause of all motions on the face of the earth,
except those of organic beings, to be treated hereafter.   It must be noticed throughout, however,
that these disintegrative effects are only directly produced upon the liquid and gaseous substances in
which the force of cohesion is very slight. Those
more solid and coherent masses, the rocks, are little
acted upon, and that only indirectly, by Liberating
Energies in the liquids and gases, as will more fully
appear hereafter.

But the Energy which thus falls upon the surface
of the earth from day to day, and sets up these separative actions in its less coherent superficial molecules,
does not long remain upon the face of our planet.
Each portion of the earth's surface is turned (on an
average) for one half of each twenty-four hours towards the sun, and for one half away from the sun.
The heat which struck it during the day and was
partly absorbed by its superficial molecules is more
or less radiated away to the ether during the succeeding night.   In such a case as that of Sahara, where
there is no organic life on whose production the
Energy is permanently expended, and little vapour of
water to retain the heat—almost all the Energy received during the daytime is radiated away at night,
so that the thermometer often sinks below the freezing point.   Here we have the naked facts uncom-

plicated by the problems of life and the interference
of rain and wind.   On the ocean, the solar Energy
absorbed by the water raises large masses of watery
vapour to a considerable height.   There, the vapour
parts sooner or later with some of its Energy to the
ether, and aggregating in the form of rain, converts
the remainder from the Potential to the Kinetic Mode,
finally yielding it up again as heat when it once more
reaches the ocean.   So in this case too, though
less conspicuously than in the former, the absorbed
Energy, after causing temporary separations, is before
long dissipated, while the particles which it affected
once more aggregate in obedience to their inherent
Forces.   On the ordinary fertile land-patches the
solar Energy is partly returned at once by radiation,
as in Sahara; partly used up in evaporation, as on
the ocean ; and partly employed in the production of
living organisms.   In the first case, the retransference
of the Energy to the ether is obvious ; in the second
case, though less immediate, it yet takes place, as explained above, when the vapour falls again as rain ;
but in the third case, the transfers are more involved,
and will have to be treated in separate chapters.
It will be enough for the present to point out that
every organism sooner or later dies, and that then the
Energy which was embodied in its production is
once more given up to the ether on the chemical

I

aggregation of oxygen and other decomposing agents with its component atoms.

Let us now look in detail at a few of the ways in which the separation, yielded up to the ether by particles of solar matter as they aggregated, is reconverted into separation between slightly-coherent mundane particles, and is finally transferred again to ether.

A lake in the northernmost part of the temperate zone is frozen over during the winter. The comparatively small amount of solar Energy which affects it does not suffice to separate its particles from their cohesive union. But when the earth shifts its position by oscillating slightly on a particular axis, the conditions of aërial refraction are altered, and the amount of radiant Energy which is concentrated on this particular spot is largely increased. The first effect of this Energy is to loosen the aggregated molecules from their solid state and to make them assume the liquid form. The Energy thus absorbed remains in the water as ' latent heat,' in other words either as Potential Energy of separation or as Kinetic Energy of motion : and when the water again freezes, it is yielded up to the surrounding atmosphere, often in the visible form of warm mist. After the separative Power has produced this first effect in melting the ice, fresh quantities are from day to day poured upon the surface of the now liquid lake. Here, the

heat produces further separation between the superficial molecules, so that even the slight cohesive power
of liquids is overcome, and the particles assume the
vaporous state. Thereupon they are raised into the
air, and drifted about by the winds, which themselves
result from the separative action of heat. After a
time, the particles lose by radiation and convection
much of their Kinetic Energy, and begin once more to
aggregate into rain-drops. These fall upon the surrounding heights, and finally find their way again into
the lake. But the Energy which raised them has by
this time been dissipated, and fresh Energy will be required to make them once more assume the form of
vapour. Nor is this all. As the drops fall upon
the mountain side, they employ part of their Energy
in overcoming the cohesion of its molecules. In this
way they slowly wear away the elevations on the
earth's surface, and carry down their particles to the
valleys or the sea. In so doing, they act as liberating
agents for the Potential Energy of these masses, and
so assist in working out the general plan of aggregation. It is true that new mountains are from time to
time slowly upheaved in place of the old ones, but
these are themselves mere apparent exceptions, as
they really represent a general lapse of the surrounding crust : and their heights are in turn worn down
by watercourses, glaciers, and percolation. In short,

the solar Energy expended in evaporation is ultimately employed as a liberating agency for the Potential Energy of separation possessed by such masses
as are raised above the general gravitative sea level
of our planet. These masses, though their cohesion
is for a while destroyed, aggregate in the end as new
sedimentary deposits ; and so the whole process becomes merely one more step in the gradual aggregation of matter and dissipation of Energy to the
ether.

Winds and storms act in similar ways. They all
arise from some kind of separation, produced in air
or water by heat ; or from the subsequent cooling of
the heated masses. In the first case, we see the absorption of separative Power ; in the second case the
re-establishment of equilibrium on its disengagement.
They, too, act as Liberating Energies for the Potential
Energy of masses elevated above the general gravitative level, as when they blow down trees, walls, or
stones, and beat the waves against a cliff. In one way
or another, every Energy which falls upon our earth
from the sun is employed in wearing down all inequalities of surface,—that is, in liberating masses
possessed of Potential Energy, and permitting them
to obey their gravitative impulses.

The special case of lightning demands a brief explanation. Throughout, we have dealt lightly with

electrical phenomena, and we must do so here once
more. The Potential Energy of the separative electricities in the thunder-cloud and the earth is in some
way a product of solar Energy. So long as they remain
apart, there is some kind of statical separation between
unknown units generally aggregated. At last, some
Liberating Energy in the shape of wind or heat brings
the charged masses within range of their mutual affinities. At once a discharge takes place, and the Potential Energy is liberated as Light, Heat, and Sound;
all of which are finally turned loose upon the ether as
radiant Energy, to pulse perhaps for ever, through the
interstellar spaces. The only peculiarity of the case
is the conspicuous and instantaneous way in which
the Potential Energy is liberated and assumes the
Kinetic Mode.

So, too, with many human machines. Organic
phenomena will demand careful separate treatment;
and until this has been given we cannot properly understand such a case as that of a steam-engine, where
the prime Energy is derived from organic products
like coal and wood. But certain simpler machines
like water-mills and windmills may conveniently be
explained at the present stage. The water which
falls from clouds on an elevated patch of ground still
possesses Potential Energy in virtue of its separation
from the general gravitative level, and as the force of

gravitation is very little interfered with by cohesion in the case of liquids, the water is enabled to form into a stream, and run down to the sea. On the way, under ordinary circumstances, it parts with most of its Potential Energy by friction, or yields it up in falling as heat. But where a considerable fall occurs, it is possible to employ this energy in turning a wheel. The wheel, being connected with other wheels and grindstones, gives up the Kinetic Energy thus derived, partly in producing separation, in opposition to cohesion, among the molecules of corn, and partly in heat or friction. The heat is of course radiated off, and the rest of the Energy remains Potential in the flour. So also with a windmill. Here the Kinetic Energy of wind, itself derived from solar rays, is transferred to the vans of the mill, and is finally used up in producing separation in the corn, or in heating the bearings and grindstones. In both cases we see, as usual, an intermediate employment of Energy for the purpose of separating material particles, but a final loss of energy from matter to the ethereal medium.

In all these cases we deal with phenomena essentially unconnected with organic life : for although the machines mentioned above are of human construction, yet, when once set in action, they can go on without human intervention until the loss sustained by friction makes their working impossible. In the next chapter we shall consider the more involved case of living organisms. Before doing so, however, it will be well to sum up the conclusions at which we have arrived regarding the general dynamical phenomena of our planet.

The earth is a proximately spherical mass of matter, held together by its own gravitation, and bulging slightly towards its equator, where its axial Energy produces the greatest effect. It revolves round the sun in virtue of its orbital motion, and it possesses Potential Energy by reason of its separate condition. This Potential Energy, however, cannot assume the Kinetic Mode, because the solar gravitation is opposed by the orbital Energy of the planet. Though the earth thus possesses two proper molar motions of its own—the axial and the orbital— its Molecular Energy has been radiated away into space from the surface at least, only the interior portion being still in a highly heated state. The interference of cohesion in this outer solid shell with the general gravitation whose Force comes into free play as the internal mass cools and contracts, gives rise to a state of tension, finally resulting in cracks and corrugations on the surface. If no external Energy intervened, the outer shell would present one uniform cold and probably solid surface, broken up into ice-clad mountains and valleys. But a fraction of the

Energy radiated into space by the aggregating masses of the central sun falls on the outer shell and there interferes with the aggregative process by setting up temporary separative action among the less coherent molecules. It keeps the atmosphere and the ocean in the gaseous and liquid forms respectively. It produces such an expansion of the equatorial air as gives rise to monsoons ; and elsewhere it heats the atmosphere of deserts, valleys, and low-lying plains so as to cause local winds and storms. It also lifts up great masses of water, which float in the air as clouds, and finally fall as rain when their Energy is dissipated. It heats the equatorial oceans, and thus rendering them lighter sets up warm ocean currents, while gravitation, drawing down the colder masses, produces the compensating cold streams. The separative nature of all these processes will be obvious when we reflect that every one of them depends upon such an absorption of radiant heat as overcomes the aggregative Force of cohesion.   But these changes are never permanent. The Energy thus absorbed is soon radiated off to the cooler ether in those less energetic periods which we know as night and winter. Unless every day and every summer new Energy were poured upon the earth to set up similar separative actions, the effects of each Energy-absorbing period would soon pass away. The vapour and the water would part with

their heat, condense, and finally freeze : while the air
would cool down, settle into stable equilibrium, and
perhaps aggregate at last into the solid state.   Moreover, the Energy which thus falls upon the earth
acts indirectly as a liberating agent to those more
solid masses which are prevented by cohesion from
aggregating in the stablest possible manner with the
general body of the planet.   By wearing down mountain sides ; by water-action, percolation, glacier-grind-
ing, and attrition of rolling bodies ; by blowing over
stones, chimneys, and trees ; by wasting cliffs, head-
lands, and river-banks ; by grinding down pebbles,
shells and refuse ; and by depositing all the débris thus
resulting in new and lower strata of mud and sand—
by all these ways and countless others, to which every
gorge, ravine, denudation valley, and seaward cliff
bears witness, the Energy poured down upon us from
the sun acts as a liberating agency to reduce the inequalities of our planet's surface, and bring every
body ultimately into closer and more intimate aggregation with the general mass.

Thus we see that on the surface of our earth
the universal process of aggregation continues in
spite of partial interruptions.   Incident Energy let
loose from the aggregating sun produces local and
temporary separations among its material particles ;
but such separations do not interfere in the end with

the general aggregating process, which they rather indirectly assist.   As elsewhere, we find all the matter engaged in a continuous course of aggregation, and all the Energy thus liberated continuously handed over to the ethereal medium.

CHAPTER V.

THE interferences caused by incident solar Energy in the aggregative processes of our earth which were considered in the last chapter mostly consisted in separative actions opposed to the molecular Force of cohesion, and, less directly, to the molar Force of gravitation. Those phenomena which we have to consider in the present chapter are the result of interferences by solar Energy opposed to the atomic Force of Chemical Affinity.

It is not here asserted that *all* the cases where solar Energy interferes with and opposes Chemical Affinity are concerned with vital phenomena. But vital phenomena form the principal instance of such interferences, and all the others may be omitted as illustrating no new principle and suggesting no new difficulty.

Regarded in their naked dynamical aspect these phenomena may be briefly described as fol-

lows.   The incident solar Energy,—besides falling
upon molecules in the slightly aggregated cohesive
states which we know as the liquid and the gaseous,
and overcoming their very moderate cohesion so as to
produce evaporation and expansion—also falls upon
certain atoms aggregated together by the Force of
Chemical Affinity, and sets up in them separative
actions, which result in the severance of these atoms
from their affinities, and the rebuilding of some
among them into those peculiar forms which may be
described as Energetic Compounds (hydro-carbons,
&c.), while the remainder are cast in a free state
upon the atmosphere.   The radiant Energy thus
employed is used up for the time being in producing
these separations, and is retained partly by the freed
elements, and partly by the Energetic Compounds,
either in the Potential Mode or in the Kinetic, or
partly in one and partly in the other (for on this
point we have as yet no certain knowledge).   The
Energy thus absorbed by the Energetic Compounds
apparently remains within them permanently, until
some incident Energy, acting as a liberating agent,
causes their atoms once more to unite with those
other free atoms in the atmosphere for which they
have affinities.   When they reunite, all the Energy
which was absorbed in producing their separation
is liberated once more by the act of aggregation,

and · is yielded up to the ether as low-grade
Energy.   While the Energy is retained by the freed
element and the Energetic Compound we may either
suppose that it is all Potential and consists merely in
the statical separation of their atoms,—on which supposition it will be exactly analogous to the case of a
rock, raised to a height and then supported so that
it cannot fall without the intervention of a liberating
Energy : or we may suppose that it is partly Potential
and partly Kinetic, and consists not only in the statical separation of the atoms, but also in a relative
motion of the atoms in the Energetic Compound,—on
which supposition it would be analogous to the case
where a collection of bodies like the solar system,
having relative motions of their own, possess Potential
Energy with reference to some other external body,
like the star in Hercules, towards which the solar
system is supposed to be moving.   It is clear that on
the first supposition the amount of Energy liberated
by the reaggregation of the atoms will be equivalent
to the Potential Energy of their statical separation :
but on the second supposition the amount liberated
will be equivalent to that Potential Energy, *plus* the
Kinetic Energy of the relative motions possessed by
the several atoms—just as, if the sun were to aggregate with any fixed star after all his planets had
already dissipated the Kinetic Energy of their several

orbital motions, and united with his mass, the Energy liberated by the aggregation would be the equivalent of the statical separation previously existing between the sun and that star; whereas, if the aggregation were to take place to-day, the amount of Energy liberated would be equivalent to the statical separation of the two systems, *plus* the Energy liberated by the stoppage of orbital and axial motion in each of the planets and satellites. It is not improbable that, in certain instances at least, we may be induced to accept the second of these two suppositions.

Translated into concrete language, the above abstract propositions may thus be more simply expressed. Solar Energy falls upon a crust containing the molecules of water, carbonic anhydride, the various nitrates in a state of solution, and other raw materials of organic matter. It finds their atoms in a condition of relatively stable chemical combination —in other words, closely bound up with one another by the Force of Chemical Affinity. Being absorbed by some or all of these atoms, it sets them free from their stable unions, by producing motions which take them beyond the sphere of their mutual attractions. It leaves the oxygen of the carbonic anhydride in a free state, while it builds up the carbon with the hydrogen of water into certain Energetic Compounds, such as starches, &c. The Energy of these compounds may be all Potential—that is to say, may consist in the fact of their statical separation from the attracting oxygen and their loose chemical apposition; or it may be partly Kinetic as well—that is to say, may also consist in the fact that the various atoms have relative movements like those of a planetary system. Furthermore, in the case of the Energetic nitrogenous compounds there is reason to suppose that a suppressed Energy is also involved. Once these Energetic Compounds have been built up, they remain permanently inert, retaining their Energy themselves in a dormant state—at least so far as human observation can detect—until some Liberating Energy brings them once more under the influence of Chemical Affinity. Thus a piece of wood or a lump of fat, once produced, remains inert, at least to outward appearance, so long as it is kept at a low temperature and isolated from disintegrating agents. But so soon as we apply a certain degree of heat to either, they burn away; or, in other words, unite once more with the oxygen from which they were previously separated, and yield up as they aggregate all the Energy of their separation and their relative movement (if any) in the form of Light and Heat. Moreover, there are several ways in which such a liberating agency can be set in action. It may be by human aid, and the intervention of external burning matters, as when we light a piece of wood or a candle by means of a match. Or it may be by the intervention of some animal organism, as when a worm burrows into a piece of wood and uses up its Potential Energy in the performance of his physiological functions, by causing its atoms to combine with oxygen within his body: or as when a carnivorous animal devours the fat, and so employs it in his physiological functions : or as when the animal which has deposited it, himself employs it for his own use, which case we see illustrated in the bear and other hibernating animals. Or, again, it may be by the set of external liberating agents which produce what we call decomposition : as when a tree decays slowly where it fell, under the influence of moisture and gentle heat : or when a dead animal decomposes in the sunlight :— though these latter cases are sure to be accompanied by the development of other organisms, which act as liberating agents, such as fungi, maggots, vibrios, and other like organisms. But whatever may be the means by which is brought about this recombination of the organic substances with the oxygen previously liberated from their affinity by solar Energy, there are two points which can be laid down as certain. *First*, that no such reaggregation of the separated atoms can take place without the intervention of a liberating agent, whether that liberating agent be moisture, solar light and heat, animal germs, fungus spores, or human interference : as we clearly see in the fact that to preserve an organic substance we may either desiccate it, or freeze it, or seclude it from light and heat, or from animal and vegetable germs, or secure it from being devoured by some other organism, or from the interference of human beings, who might burn it or otherwise cause its re-aggregation with oxygen : while on the contrary we know that exposure to one or other of these liberating agents will bring about such reaggregation (or de-composition, as it is oftener though less accurately called) in every kind of organic matter. *Second*, that on the whole and in the vast majority of cases almost every piece of organic matter aggregates at last with the oxygen or other free atoms from which its ele-ments were at first severed, and yields up its Energy to the ether in some more or less conspicuous manner. Thus, sooner or later, every plant, if left to itself, dies and decays : that is, recombines with oxygen slowly, under the influence of moisture, light, and heat, and yields up its Energy by inconspicuous degrees ; while every animal, if left to itself, similarly dies and decays, probably under the influence of other small animal germs, which use up its contained Energies in carry-ing on their own activities : and so, in both these cases, the atoms finally reaggregate in stable com-

K

bination, while the Energy is yielded up, immediately perhaps to surrounding matter, but finally to the ethereal medium. So, too, if the plant or animal is devoured by an animal organism, its atoms are made to combine with oxygen within the devouring organism, and their Energy is yielded up as heat and as movement, either of internal parts or of external limbs, and is thus finally dissipated. And even if, as in the case of peat, petroleum, and coal, or of the Siberian mammoths, the Energetic Compounds are long secluded by their circumstances from Liberating Energies, it may yet finally happen that human activity may intervene to liberate their Energies, as we see when we burn coal, petroleum, or peat, or when we exhume mammoths, and so expose them to the decomposing (liberating) action of the sun and organic germs. So that organic life, when closely considered, proves dynamically to be a mere special case of the general laws : and we see that though it is in its nature separative, as being the product of solar Energy absorbed for a time by particular mundane particles, it nevertheless results in a final reaggregation of atoms in stable combination, and dissipation of Energy to the ethereal medium.

A word of explanation is necessary. It may be asked, why will not the organic compounds aggregate at once with the free oxygen, and why do they need the liberating agency of heat or other Energy? The answer is probably analogous to that which we gave in the case of cohesion. Unless the atoms are brought very close to one another they cannot apparently get within the range of their mutual affinities, and mere mechanical juxtaposition is insufficient for this purpose without such atomic vibration as will bring them into close quarters with one another. But the more complex animal compounds, as we shall see hereafter, seem to possess high Kinetic Energy of their own, which can only be kept up in the circumstances of the body: and it is probable that they spontaneously decompose (or split up into simpler and less Energetic compounds) with a liberation of Energy on any direct contact with external agencies.[1]

In the present work no attempt will be made to account for the origin and development of living organisms. That task has been satisfactorily performed in portions by Darwin, Haeckel, Müller, Huxley, Wallace, Hooker, and others, while a more comprehensive and systematic view of the whole process has been given by Mr. Herbert Spencer (whose name I can never pass by without the expression of my deepest intellectual gratitude and veneration). Their results can easily be translated into terms of the theory advocated in this work: and they

---

[1] About this point the author is now extremely doubtful.

have not sufficiently direct dynamical bearings to concern us greatly in our present inquiry.[1] It must suffice here to recognise the fact that life owes its origin to the chemically-separative action of ethereal undulations on the cooled surface of the earth, especially carbonic anhydride and water, and that the existing diversity of organic forms is due to the minute interaction of dynamical laws.

It will, however, be desirable to point out that life is essentially separative in its nature, because the identity of Energy with separation is the main point insisted upon in the present treatise, and life is the Mode of Energy with which human beings are most familiar, and from which they form their conception of all its other modes.

Life, then, is shown to be essentially separative, first, because it is a product of solar Energy, acting upon the superficial matter of the earth. This Energy is the locomotive form of the statical separation once existing between the particles of the sun's mass. When it falls upon the earth, being then in the ethereal form, we know that it is partly absorbed by various loosely aggregated superficial material molecules, in

---

[1] It must be understood that no disrespect towards such inquiries is intended in the present passage. On the contrary, there can be no doubt that, bearing as they do on all our acts and theories as living beings, these questions are of paramount practical importance. But they are not necessary to the present subject, and they have already been treated sufficiently by proper and competent authorities.

which it sets up separations that overcome the molecular Force of cohesion, and so produces winds, storms, ocean currents, clouds, &c. Now it similarly falls upon certain other molecules, among whose atoms it sets up separations that overcome the atomic Force of Chemical Affinity, and so produces starch, albuminoids, free oxygen, and other like chemically Energetic bodies. The separative nature of this process is obvious. Without the disjunctive solar Energy there could be no life, just as there could be no wind, ocean currents, rain, or clouds. All the stable chemical compounds would remain for ever in the aggregated state, unless the solar Energy came in to separate them. Again, life is seen to be essentially separative by its mechanical position and effects. Trees, plants, and animals stand out for the most part at a visible elevation from the mass of the earth's solid crust, and when they die, large portions of them fall down and are reaggregated with its substance. The heat which sets up evaporation in leaves causes a capillary circulation in the vessels and cells of the plant, whereby water, holding in solution nitrogenous salts and mineral matters generally, is raised to every part of its surface; and then a large portion of this water is evaporated, while the mineral matters remain in the leaves and fibres. In all this we obviously see separative action opposed to gravitation, as above we saw it opposed to chemical affinity. Still more clear is this point of view in animals, which climb trees, plants, rocks, and mountains; which fly to great heights in the air; and some of which carry about great masses of bone, while others lift stone and brick to conspicuous elevation as houses, towers, and steeples. No one of these separative acts could have been performed without the intervention of solar Energy. But it is especially in its reconversion that organic matter shows its separative nature. As its atoms reaggregate, they give out heat, which obviously causes molecular separation in the surrounding bodies. The animal organism is perpetually in such a heated condition, and is perpetually parting with heat which goes off to swell the volume of ethereal Energy. So that in every way life reveals itself as an effect of the separative action exerted by ethereal Energy on the superficial material particles of our planet.

Succeeding chapters will deal with the phenomena of vegetal and animal life severally, as enforcing and illustrating these principles. For the present we may content ourselves with a brief summary of the results already attained.

Organic life is one of the effects wrought by incident solar Energy on the surface of the earth. It originates mainly in separative actions, whereby

atoms are severed from relatively stable chemical
combinations, and are either turned loose upon the
atmosphere in a free state, or are built up into Energetic Compounds. But through the action of liberating agents, also of solar origin, these free atoms and
Energetic Compounds for the most part sooner or
later recombine; whereupon the absorbed Energy is
once more liberated and turned loose upon the ether.
Organic life is thus a transitory result of the general
aggregating process during which the Energy liberated by the aggregation of particles in one mass
falls upon the aggregated surface of another mass,
and there sets up separative actions, which, however,
are most often only temporary in their effects, owing
to the subsequent incidence of Liberating Energies,
whereby the absorbed Energy is once more turned
loose upon the ether.

CHAPTER VI.

THE VEGETAL ORGANISM.

ALTHOUGH in the last chapter, where we treated of life generally as a product of incident solar Energy, we made little distinction between the two main forms of life, it must yet be understood that the relation which, as wholes, they bear to the incident Energy is exactly contrary. Vegetal organisms, as a rule, are accumulators of Energy, and not expenders : animal organisms, as a rule, are expenders of Energy and not accumulators. In other words, the vegetal organism is a case where incident Kinetic Energy is setting up separative actions between aggregated atoms, and is being absorbed (or potentialised) in the separation so produced : while the animal organism is a case in which the atoms so separated are being reaggregated, and their Energies, Potential or Suppressed, are assuming the Kinetic Mode, either as heat or as visible motion. The energy absorbed and potentialised by the plant, is kineticised and given off by the animal.

These statements must only be accepted as true in the gross, and with certain deductions duly noted hereafter.

The plant is the origin of all the Energy possessed by all living beings. The separation between the atoms of water, carbonic anhydride, and nitrogenous salts, which takes place in its tissues under the influence of sunlight, is the Potential Energy which becomes Kinetic in the growing seed, the expanding flower, and the leaping or flying animal. We may therefore briefly trace the life-history of a plant, as throwing some light upon the dynamical nature of life generally.

Every plant starts as a spore or seed, cast off from a previously existing plant. This first germ contains some small materials for growth for the young plant in the form of Energetic Compounds, whose Potential Energy is to become Kinetic in the act of germination. In order, however, to produce this effect, liberating agents are needed; and these liberating agents are generally three in number, moisture, heat, and light. These, acting upon the materials in the seed, either cause them to aggregate with other matters, or overcome the suppressing Force; and in consequence the materials yield up their Potential or Suppressed Energies in that determinate form imposed by the specific conditions and known as ger-

mination. The amount of Energetic materials supplied
to the new plant (or the fresh year's growth) may be
very great, as in the potato tuber, the lily bulb, and
the wheat grain, or it may be very little, as in fungi,
ferns, and cryptogams generally: but in every case,
if the plant is to continue living, there must be enough
Energy to permit of its assuming the shape in which
it can begin to be acted upon by the sunlight, and to
assimilate fresh matter under the influence of that
incident Energy.   This stage is reached when leaves
are produced.   On the surface of these leaves the
solar Energy produces evaporation, and this evaporation gives rise to a general capillary action, whereby
water is raised into the leaves.

In these leaves the sunlight, acting upon carbonic
anhydride sucked in from the atmosphere, frees the
carbon atoms from their union with the oxygen, and
builds them up with the hydrogen into hydrocarbons
—Energetic compounds: while the oxygen is turned
out upon the atmosphere in a free state.   Nitrogenous
salts in solution have also been supplied by the water,
and from these and the starch, the plant in some unknown way builds up the protoplasm which forms the
moving portion of all living organisms.   The starch,
sugar, albuminoids, and other organic compounds
thus produced are then circulated all over the plant,
where they supply the materials for growth, and

develop new leaves, which in turn increase the
amount of Energetic matter in the plant. Part of
the Energy thus absorbed is used up by the plant
itself in its own physiological processes. The growth
of each cell doubtless involves the expenditure of
Energy—that is to say, some Energy previously contained by the protoplasm assumes thereupon the
Kinetic Mode, and is in part yielded up to the ether.
In the larger physiological processes, such as germination or inflorescence, it is certain that such dissipation
of Energy takes place, in the first place because free
oxygen is absorbed and carbonic anhydride is evolved,
which shows that some of the contained carbon has
reaggregated with the oxygen ; and in the second
place because a rise of temperature can be shown to
accompany these processes. Accordingly we may
conclude that the motions which take place in plants
are due to the reaggregation of certain Energetic
particles with the free atoms of their neighbourhood,
and that while some of the Energy thereupon liberated
has assumed the form of Molar Motion, part of it has
been dissipated as heat. But a large portion of the
Energy remains dormant in the plant, either in the
molar or the atomic species. The leaves and stem as
wholes, viewed mechanically, represent the former: the
starch, protoplasm, and wood, viewed chemically, represent the latter. When the plant dies or is devoured,

on the average of instances the greater part of this
Energy is rendered Kinetic, and ultimately yielded up
to the ether.   Take first the case of a tree which
dies a natural death. At the end of each year its
leaves fall. Before they do so, they render up their
most important chemically Energetic products to
the permanent portions of the complex organism ; but
inasmuch as they will burn, they retain a certain
amount of atomic Energy in their cellulose ; and inasmuch as they are elevated above the general level,
they possess Molar Potential Energy in their position.
When the leaves drop off by the weakening of cohesion
at their bases (along a pre-arranged line) their Molar
Potential Energy becomes Kinetic in the act of falling, and is dissipated as they reach the ground. The
dead leaves, lying on the earth, now consist mainly
of inorganic earthy matter and cellulose.   The
action of moisture, heat, and light, as liberating
agents, soon sets up decomposition : and the mineral
matter lies *in situ*, while the organic substances combine with the surrounding oxygen. When the whole
tree dies the same process is repeated on a larger
scale. The actions of moisture, heat, and light, combined with those of fungi, worms, &c., are liberating
agencies which cause the trunk to decay and fall, and
afterwards produce more or less complete decomposition of the whole tree as it lies. In a few cases,

which will be treated of hereafter, the stored-up
Energies are long retained in coal, peat, and vegetal
débris : but on the average of instances, almost all the
Energy absorbed during one year has been yielded up
by the next.   When the plant is devoured by an
animal or burnt by man, it undergoes a somewhat
different yet ultimately identical cycle of changes,
which will be more fully detailed in our next chapter :
for the present it will suffice to say that its Energetic
Compounds combine with free oxygen within the
animal organism, or the fireplace, and that their
Energy is used up in the production of motion and
heat, and is thus, as usual, finally transferred to the
ethereal medium.

The special case of reproduction requires a few
additional remarks.   Where this function is carried
on by inflorescence, we have a series of leaves produced which are expenders of Energy, instead of
being accumulators, growing and unfolding by the
employment of Energy stored up in other parts of
the plant.   Most especially is this the case with the
pollen, ovule, and seed.   In the protrusion of the
pollen-tubes and the growth of the embryo, we see
conspicuous instances of the employment and dissipation of previously stored Energy.   In the developed
seed we sometimes find a store of albumen ; and in
any case we have in the embryo itself a nitrogenous

mass which is able, under the influence of moisture and heat (liberating agents), to aggregate in part with oxygen and produce germination.   Somewhat similar in their dynamical nature are those morphologically unlike propagating portions which lay up nutriment for the future growth of the individual or its offspring.   Such are the roots and tubers of potatoes and beets, the tubers of the orchis and dahlia, the corms of the saffron, and the bulbs or bulbils of the onion and the tiger-lily.   In every case, motion in plants is caused by the aggregation of free oxygen with the Energetic Compounds of the plant, and by the employment of the Energy thus liberated for the production of Molar motion.

It will thus be seen that even those plants which are on the whole accumulators and storers of Energy are themselves to some extent likewise expenders of Energy: and that the broad distinction which we have drawn between the vegetal and the animal organisms, viewed dynamically, must not be pressed too close.   In growth, in inflorescence, and in germination, the plant is essentially an animal. It is only in assimilation that it displays the characteristic vegetal function of transferring Energy from ether to material particles mainly by the production of hydrocarbons, *plus* free oxygen, from carbonic anhydride and water.   We may thus say roughly, in

reference to our present standpoint, that the assimilating leaf, frond, or thallus, is the only true
plant. Nor is this all. Many organisms, classed
morphologically as plants, are in their dynamical
aspect the analogues of animals : that is to say, their
functions are wholly expensive of Energy and not at
all accumulative.[1] The leafless parasites (orobanche,
cytinus &c.) fasten upon some other plant, and
without themselves contributing to the general store
of Energy, employ the Energetic Compounds laid up
by their host, in the production of their own flowers
and seeds. A much larger and more important class
is that of fungi, which live upon the roots, stems,
seeds, or tubers of other plants, upon the bodies or
the dung of animals, or upon the generally diffused
undecomposed organic matter of the soil. But
whencesoever they derive their materials, they always
thrive upon previously-composed Energetic Compounds, whose Energy they liberate with almost
explosive power. They are like animals in never
accumulating Energy, while expending that which has
been previously accumulated by other plants. It
is noticeable that all these quasi-animal functions can
be carried on in the absence of light, that is, of high-

---

[1] Allusion is not here made to insectivorous species, like Dionæa,
Nepenthes, and Drosera; but to plants which derive their whole
material from previously organised matter.

power radiant Energy. Thus, a seed will germinate, a hyacinth will grow from its bulb and produce blossoms, a potato will sprout from its tuber, a flower will open, and a fungus will pass its whole life, under proper conditions of heat and moisture combined with the presence of oxygen, in a perfectly dark cellar: because the Energetic compounds, and the free oxygen whose aggregation liberates their Energy, are all stored up in the plant or its environment beforehand. But no assimilation, no separation of atoms from their stable unions, can take place except under the disjunctive influence of radiant Energy.

So, in spite of these numerous exceptions—these quasi-animal functions of all plants, and these large groups of plants with none but quasi-animal functions—the distinction which we have marked between plants and animals is yet of cardinal importance, and for this reason. Though some plants are quasi-animal, no animal is quasi-vegetal.[1] All the Energetic Compounds which enter into the composition of any living organism are derived, directly or indirectly, from plants. In the leaf or thallus or body of some plant or protophyte all the organised materials have taken their rise, under the separative influence of radiant Energy.

---

[1] Exception may be made of a few doubtful chlorophyll-containing animals.

To sum up, the conclusions at which we have arrived are these. Solar Energy, playing upon certain superficial material particles of our planet, separates their atoms into Energetic Compounds and free elements. The masses immediately built up of these Energetic Compounds, together with certain inorganic (or stably-compounded) substances, are known as plants. They go on continuously assisting (by means of their chlorophyll) in the similar separation of other atoms by solar Energy, some of which (hydrocarbons) swell their mass, while others (oxygen) are turned loose upon the atmosphere. The Energy thus stored in the matter of the plants and the free elements about them, does not remain perpetually connected with the same particles. Partly it is used up in the physiological operations of the plant : partly it is stored away in seed, tubers, &c., for future physiological operations : partly it is dissipated at the death of the plant. In a vast number of instances the plant is eaten by an animal, and in that case the reaggregation of elements and dissipation of Energy takes place within the animal's body. So that, in the majority of instances, the Energy radiated from the sun into the ether, and temporarily employed on the surface of our planet in the production of vegetal life, is sooner or later cast once more upon the ether, to make its way for ever through the interstellar spaces.

L

Only a small portion remains here, dormant in wood, coal, and peat ; and even that small portion, as we shall hereafter see, is finally used up by animals (including man) for some purpose connected with their vital necessities.

CHAPTER VII.

THE ANIMAL ORGANISM.

In the last chapter we saw roughly what were the dynamical relations of those organisms which act mainly as accumulators of Energy. In the present one we must make a similar investigation regarding the dynamical relations of those organisms which act mainly as expenders or dissipators of Energy.

Amongst these, · as already noticed, are many themselves of vegetal origin ; and it may simplify matters if we first look briefly at their nature, afterwards noting the principal points of distinction between them and the animal organism proper.

A fungus grows upon a decaying tree. It has its origin in a spore ; and this spore, alighting in the neighbourhood of previously-accumulated Energetic Compounds, has its own Energies liberated by heat and moisture ; and thereupon becomes in turn a liberator of the Energies in the organised matter around it. These it gathers into its mass, and gradually dissipates, with the exception of that portion which it bequeaths to *its* spores, thereby once more beginning a similar cycle of changes. Wherever the fungus seats itself,—whether on the root or stem of a plant; or on a seed, tuber, or other receptacle of Energetic Compounds destined for future growth; or on an animal body; or on a patch of soil containing dispersed undecomposed organic matter;—it adds nothing to the total of Energy, being merely a dissipator of the Energy already accumulated.

Now, the animal organism is a mechanism in which these same processes take place, but take place much more rapidly and conspicuously, and are accompanied by certain secondary phenomena. As the highest organisms show all the processes of the lower, and also some others peculiar to themselves, it may be convenient to take an example from the upper ranks of animal life to illustrate the specialities of the case.

A young vertebrate begins its existence as a small mass of Energetic Compounds presenting elementary organisation, surrounded by another mass (more or less) of comparatively unorganised Energetic material. As in the case of the plant, the material may differ in amount, but must be sufficient, under the liberating agency of heat, to carry on the process of organisation to such a point that the young organism can obtain the necessary further material for itself. In the case of a bird, this material is supplied by the food-yolk ; in that of a mammal it is supplemented by nutriment derived from the maternal circulating system. But in every case, the young plant and the young animal are alike in this, that each begins its life as an ex-pender of previously-accumulated Energy. It is needless to add that the presence of free oxygen, which combines with the food-stuffs to produce carbonic anhydride, and so yields up its Energy for the act of organisation, is in both cases indispen-sable. So soon, however, as the self-sustaining de-gree of organisation has been reached, a wide diffe-rence begins to manifest itself. The plant spreads its leaves to the sun and the air, and drinks in carbonic anhydride and water, from which the ethereal Energy separates part of their oxygen, and manufactures starch and other organic compounds. The animal, on the contrary, devours the compounds thus formed, and drinking in the free oxygen, causes them to re-aggregate within his body, using up the Energy so liberated, partly for the production of heat, partly for physiological processes, and partly for locomotion. What may be the exact nature of these conversions we do not fully know ; and even if we did, they could only be detailed in a complete work on Physiology : but it is sufficient for our purpose to point out that

Energetic carbonaceous and nitrogenous matters pass into the body by one channel, and free oxygen by another ; that they leave the body as carbonic anhydride, ammonia, water, and other de-energised products ; and that heat and motion have been given out meanwhile. The animal organism is thus essentially a seat for the reaggregation of matter and the dissipation of Energy.   It is, however, probable that part of the Energy thus liberated is not immediately dissipated, but is used up for the time being in the so-called synthetic processes of the body.   What these really are, we do not thoroughly understand : but it seems likely that they may be combinations of many atoms, possessing high relative Kinetic Energy, upon whose decomposition the Kinetic Energy is liberated.[1]   Thus, a manifestation of Energy accompanies the conversion of sugar into lactic acid, or into alcohol and carbonic anhydride, although no new oxygen is united during the process.   At any rate, waiving all speculation, it is certain that these bodies, unlike ordinary compounds, possess Energy in their composite form, which is dissipated when they separate into simpler bodies.

While the animal lives, he is perpetually taking into his organism Energetic Compounds stored up in plants, or temporarily deposited in the tissues of other

[1] This the author now greatly doubts.

animals, and using up their Energies for his own activities. In some cases the matters thus absorbed are immediately employed for physiological processes : but in other cases they are stored up, like the starch and albumen of seeds or tubers, for future use. A bear living through the winter on his own fat, or a camel consuming his humps during a journey, is the exact analogue of the sprouting wheat and of the potato or hyacinth grown in a cellar. When the animal dies, heat and other animal germs act as liberators for his stored-up tissue Energies ; and decomposition rapidly sets in, resulting in the final formation of stable compounds. Thus the matter which during the animal's life possessed Energy of chemical separation in its atomic composition,—Energy of molar separation in its erect position and frequent elevation in the air or on mountain tops,—Energy of molecular motion in its heat,—and Energy of molar motion in its locomotive processes,—becomes at last a number of chemically stable masses, partly aggregated with the earth's surface, and partly floating as carbonic anhydride, incapable of resuming its separate and self-moving condition except by the intervention of fresh solar Energy through the agency of vegetal life.

Viewed from a wide standpoint, we may say that animals act as liberating agents for the Energies stored up in plants. They are therefore links in that general chain of processes whereby separate portions of matter are made to aggregate in the stablest union, and their previously-existing separation is imparted to the ether.

CHAPTER VIII.

GENERAL VIEW OF MUNDANE ENERGIES.

WE have now completed our brief survey of the cosmical facts known to us at present, and examined their
congruity with our general theory of two opposing
Powers, aggregative and separative. But before we
close the subject it may be well to look briefly at the
facts of mundane Energy in their entirety, with especial
reference to the part played by man.

By the term 'Cosmical Energies of the Earth' we
may understand all that Energy which our planet
possesses in virtue of its statical separation from the
sun and the other sidereal bodies. By the term
'Proper Energies of the Earth' we may designate all
that Energy which the material particles composing
the earth's substance now possess or formerly possessed in virtue of their own original separation from
one another, as masses, molecules, or atoms. Finally,
by the term 'Derived Energies of the Earth,' we may
understand all that Energy which our planet has ab-

sorbed from the radiated Energy of other aggregating masses elsewhere : and as the amount of such absorbed Energy derived from the ' fixed' stars or reflected from the moon and planets is practically without conspicuous effects on the earth's surface, we may consider this term as equivalent to directly incident solar Energy.

The Cosmical Energies need not long detain us. Doubtless, as the earth loses orbital Energy by ethereal friction, it is slowly approaching the sun, while the sun in turn is approaching its own central point of attraction ; but these remote possibilities possess for man only a speculative interest, and have no influence on practical mechanical activities.

The Proper Energies of the Earth are more interesting to humanity.   (1) First, come the facts of the planet's orbital Energy and nutation, which indirectly yield the phenomena of winter and summer. (2) Next comes its axial motion (the indirect cause of day and night),[1] of whose dissipation the tides are a concomitant.   (3) More purely terrestrial are the phenomena of earthquakes, volcanoes, &c., which are Kinetic transformations of the statical separation existing between the superficial and central masses.

---

[1] Indirect, because the real cause is the incident sunlight. Were there no sun, the axial energy might still continue, but not, of course, the daylight.

(4) Derived from the last-named Energy is the Potential Energy of mountains and other masses raised
by lateral pressure above the general level of the solid
and liquid surface. In both these cases,—that of the
whole cohering crust, and that of special raised masses
—small portions of the Potential Energy become from
time to time Kinetic under the influence of liberating
agencies; in the first case, we know the result as an
earthquake, in the second as a landslip. (5) Next
may be mentioned the internal heat of the earth,
small portions of which are always escaping by conduction through the cohering crust. (6) Lastly, we
may mention the Potential Energy of chemical separation in free elements, like sulphur &c., within the
earth's crust, if these ought not to be regarded as of
organic origin, and consequently included in the list
of Derived Energies. All these Proper Energies
are the surviving forms of the separation once existing between the various portions of our earth. (But
in the case of the cohering crust, the mountains, and
the free elements, the Energies remain as statical separations to our own time. In the case of the orbital
and axial Energies, the separation has assumed the
form of continuous molar motions. In the case of
the internal heat, it has assumed that of continuous
molecular motion.) And in every earthquake, landslip, falling cliff, or tumbling stone, we see these Po-

tential Energies assuming the Kinetic Mode under our
very eyes.   Nor is it necessary to show in detail
how the earth is gradually parting with all these
Proper Energies.   The orbital and axial motions
are being dissipated by ethereal friction or by the
moon's attraction.   The internal heat is being dissipated by conduction.   The Potential Energy of the
crust is being given up from time to time by earthquakes, or, after assuming the form of heat in volcanic
eruptions, is being radiated off into ether.   And the
mountains, cliffs, and other elevated portions locally
raised for awhile (to outward appearance) by these
. widespread disturbances, above the general level, are
being for ever worn down by rain, storms, roots,
animal footsteps, and other results of those Derived
Energies which we have next to examine.   So that
the remaining Proper Energies of the earth (most of
them having been long since dissipated; after the
partial aggregation of its matter, during the cooling
of its crust) are still being cast loose, in one form or
another, upon the Energy-absorbing ether ; while a
corresponding aggregation of its matter is for ever
taking place.

But the great mass of those Mundane Energies in
which man is directly interested belong to the class
of Derived Energies.   And these, as we have already seen, may be considered as practically equiva-

lent to the directly-incident solar energy and its deri-
vatives. They are difficult to classify, owing to their
rapid changes, but the following division may cast some
light on their nature. Solar Energies are either
Direct, as light and radiant heat, or Absorbed, as in
clouds, organisms, &c. The latter or Absorbed class
may be again divided into those of Inorganic Origin,
and those of Organic Origin. Each of these will demand separate treatment.

Direct Solar Energy is the radiant Energy which
from moment to moment is cast upon our planet from
the sun. If the surface of the earth were composed
of a perfectly reflecting non-absorbing substance, all
this Energy would immediately be reflected back into
space. As it is, a small portion is so reflected, but
the greater part is absorbed by various superficial
bodies in the production of motion and separation
between their parts.

Absorbed Solar Energy, employed for Inorganic
purposes, gives rise to the following among other
phenomena. (1) The Energetic gaseous condition of
the atmosphere. (2) The Energetic liquid condition
of water in temperate climates. (3) The melting of
ice. (4) The act of evaporation and raising of vapour.
(5) The production of winds and storms. (6) The
production of ocean currents. Some of these Energies are Kinetic, as in the case of the gases, liquids,

&c. : but in other instances the Energy is rendered Potential for awhile, as in the case of the floating cloud, the head of water, and the ice or snow upon the mountain tops. And, finally, these Potential Energies are constantly becoming once more Kinetic, as when the rain falls, the river flows, and the glacier or ava- lanche slides down the valley.   Each such Kinetic Energy is of course in the act of being dissipated, by friction or otherwise, to the ethereal medium.   And as before noticed, these Inorganic Absorbed Energies become liberating agents for the proper Potential Energy of the Earth, when rain or rivers wear down rocks and mountains; when glaciers or avalanches grind their bed and tear away the stones; when storms beat the waves against the cliff; and when winds upset rocky masses.   Moreover, they also act as liberating agents for Potential Energy of Organic Origin, when lightning, rain, or wind wears down and overthrows trees or buildings, when storms sink iron ships, and when avalanches overwhelm villages.

Still more difficult to trace, because of their numerous involutions, are the vicissitudes of that Energy which assumes the Organic form. Yet we must endeavour to give some account of its main phases.   The Energy which falls on the growing plant lays up Energetic Compounds in the plant's sub- stance and raises it to a position of visible height.

Part of the Potential Energy thus obtained, the plant uses up in its own processes : part remains for awhile inherent in its tissues. But, for the most part, sooner or later it is either devoured by an animal, or else dies. In the first case, the animal digests it, and uses up its Energy in its own processes as heat and motion. In the second case, fungi grow upon it, worms devour it, water disintegrates it, and in one way or another it yields up its Energy at last to ether. Besides the Energy thus stored up in existing organisms, there is some Energy belonging to extinct organisms yet remaining on our earth. The small amount which is stored up in wood, meat, Siberian mammoths, desiccated diatoms, and other like forms, may be neglected on account of its insignificant quantity. But there are considerable stores of Energetic material, known as coal, peat, rock-oil, &c., which deserve a passing mention. These are so situated that without a liberating agent they could not be dissipated. Such a liberating agent they find in man, who is so rapidly using them up that he is now beginning to look forward to a future when all such stores will be exhausted, and when he will have to depend for his stock of Energy on the immediate daily supplies from the sun. As to the animal organisms, they are themselves entirely expenders of Energy, and their whole life consists in a reaggregation of matter and

consequent dissipation of Energy. In one way, however, such organisms leave portions of their Energy
for awhile in a Potential form, namely by building.
Every house, wall, church, &c., is a mass raised to a
height by means of Energy: and it may retain its
Energy, in the absence of a liberating agent, for a
considerable time.   But in the end, Solar Energy,
in some one or other of its transformations, will act as
a liberating agent to reduce these irregularities and
wear down their masses. Either by rain, wind, fire,
roots of trees, lightning, or the hand of man, every
building sooner or later totters to the ground. And
if it escapes all these, the earth's own Proper Energy
may prove its overthrow, by tides, earthquakes,
subsidences, or volcanic eruptions.   So that, as
in every other case, we see the matter ultimately
aggregating and the Energy handed over to the all-
absorbing ether.

Thus the earth is for ever parting with its Energy
in every shape. It is slowly aggregating with the
sun and the fixed stars. It is losing its orbital and
axial motions. By internal cooling, by subsidences,
earthquakes, and volcanic eruptions, by radiation
from lava and hot springs, it is getting rid of the
proper separation and motion inherent in its own
mass.   Solar Energy falling upon it prevents and
counteracts for awhile its total aggregation,—liquefies

the ocean, keeps the atmosphere gaseous, creates winds and currents, piles ice on mountain-tops, stores up Energetic Compounds in vegetal and animal organisms, lays by coal and peat, builds castles and cathedrals, smelts iron, and separates chemical bodies in phials and jars.   But this very same Solar Energy acts as a liberating agent in the long run not only for its own Potentially-Energetic products, but also for the proper Potential Energy of the earth.   It wears down cliffs, mountains, and table-lands, it melts the ice and snow on the mountains, it sets the animal to devour the plant ; it drives man to dig and burn the coal and petroleum ; it overthrows the buildings he has piled ; it rusts his iron implements; it corrodes his chemical reagents.   In one way or another, all the Energy of the earth's own primitive separation, and all the intercepted Energy of the sun's primitive separation, go together to swell the Energy of the ether, the great waste-heap of the Universe.

The author is not disinclined to consider the ether as composed primordially of the most tenuous ponderable matter on the outside of all aggregating systems, to which therefore their Energy has been transferred, so as at last practically to neutralise the inherent Forces.

Spottiswoode & Co. Printers, New-street Square, London.

M

A Catalogue of Works

IN

GENERAL LITERATURE

PUBLISHED BY

MESSRS. LONGMANS, GREEN, & CO.

39 PATERNOSTER ROW, LONDON, E.C.

MESSRS. LONGMANS, GREEN, & CO.

Issue the undermentioned Lists of their Publications, which may be had post free on application :—

1. MONTHLY LIST OF NEW WORKS AND NEW EDITIONS.

2. QUARTERLY LIST OF ANNOUNCEMENTS AND NEW WORKS.

3. NOTES ON BOOKS ; BEING AN ANALYSIS OF THE WORKS PUBLISHED DURING EACH QUARTER.

4. CATALOGUE OF SCIENTIFIC WORKS.

5. CATALOGUE OF MEDICAL AND SURGICAL WORKS.

6. CATALOGUE OF SCHOOL BOOKS AND EDUCATIONAL WORKS.

7. CATALOGUE OF BOOKS FOR ELEMENTARY SCHOOLS AND PUPIL TEACHERS.

8. CATALOGUE OF THEOLOGICAL WORKS BY DIVINES AND MEMBERS OF THE CHURCH OF ENGLAND.

9. CATALOGUE OF WORKS IN GENERAL LITERATURE.

ABBEY and OVERTON.—The English Church in the Eighteenth Century. By CHARLES J. ABBEY and JOHN H. OVERTON. Cr. 8vo. 7s. 6d.

ABBOTT.—Hellenica. A Collection of Essays on Greek Poetry, Philosophy, History, and Religion. Edited by EVELYN ABBOTT, M.A. LL.D. Fellow and Tutor of Balliol College, Oxford. 8vo. 16s.

ABBOTT (Evelyn, M.A. LL.D.)— Works by.

A Skeleton Outline of Greek History. Chronologically Arranged. Crown 8vo. 2s. 6d.

A History of Greece. In Two Parts.
Part I.—From the Earliest Times to the Ionian Revolt. Crown 8vo. 10s. 6d.
Part II. Vol. I.—500-445 B.C. [*In the press.*
Vol. II.—[*In preparation.*]

ACLAND and RANSOME.—A Handbook in Outline of the Political History of England to 1887. Chronologically Arranged. By A. H. DYKE ACLAND, M.P. and CYRIL RANSOME, M.A. Crown 8vo. 6s.

ACTON.—Modern Cookery. By ELIZA ACTON. With 150 Woodcuts. Fcp. 8vo. 4s. 6d.

A. K. H. B.—The Essays and Contributions of. Cr. 8vo.
Autumn Holidays of a Country Parson. 3s. 6d.
Changed Aspects of Unchanged Truths. 3s. 6d.
Commonplace Philosopher. 3s. 6d.
Counsel and Comfort from a City Pulpit. 3s. 6d.
Critical Essays of a Country Parson. 3s. 6d.
East Coast Days and Memories. 3s. 6d.

[*Continued on next page.*

A

A. K. H. B.—**The Essays and Contributions of**—*continued.*

Graver Thoughts of a Country Parson. Three Series. 3*s.* 6*d.* each.

Landscapes, Churches, and Moralities. 3*s.*6*d.*

Leisure Hours in Town. 3*s.* 6*d.*

Lessons of Middle Age. 3*s.* 6*d.*

Our Little Life. Two Series. 3*s.* 6*d.* each.

Our Homely Comedy and Tragedy. 3*s.* 6*d.*

Present Day Thoughts. 3*s.* 6*d.*

Recreations of a Country Parson. Three Series. 3*s.* 6*d.* each.

Seaside Musings. 3*s.* 6*d.*

Sunday Afternoons in the Parish Church of a Scottish University City. 3*s.*.6*d.*

'To Meet the Day' through the Christian Year: being a Text of Scripture, with an Original Meditation and a Short Selection in Verse for Every Day. 4*s.* 6*d.*

American Whist, Illustrated : containing the Laws and Principles of the Game, the Analysis of the New Play and American Leads, and a Series of Hands in Diagram, and combining Whist Universal and American Whist. By G. W. P. Fcp. 8vo. 6*s.* 6*d.*

AMOS.—**A Primer of the English Constitution and Government.** By SHELDON AMOS. Crown 8vo. 6*s.*

Annual Register (The). A Review of Public Events at Home and Abroad, for the year 1889. 8vo. 18*s.*

*** Volumes of the 'Annual Register' for the years 1863-1888 can still be had.

ANSTEY.—**Works by F. Anstey,** Author of 'Vice Versâ.'

The Black Poodle, and other Stories. Crown 8vo. 2*s.* bds.; 2*s.* 6*d.* cl.

Voces Populi. Reprinted from *Punch.* With 20 Illustrations by J. BERNARD PARTRIDGE. Fcp. 4to. 5*s.*

ARISTOTLE.—**The Works of.**

The Politics, G. Bekker's Greek Text of Books I. III. IV. (VII.) with an English Translation by W. E. BOLLAND, M.A. ; and short Introductory Essays by A. LANG, M.A. Cr. 8vo. 7*s.*6*d.*

The Politics : Introductory Essays. By ANDREW LANG. (From Bolland and Lang's 'Politics.') Crown 8vo. 2*s.* 6*d.*

The Ethics ; Greek Text, illustrated with Essays and Notes. By Sir ALEXANDER GRANT, Bart. M.A. LL.D. 2 vols. 8vo. 32*s.*

The Nicomachean Ethics, Newly Translated into English. By ROBERT WILLIAMS, Barrister-at-Law. Crown 8vo. 7*s.* 6*d.*

ARMSTRONG (G. F. SAVAGE-) — **Works by.**

Poems : Lyrical and Dramatic. Fcp. 8vo. 6*s.*

King Saul. (The Tragedy of Israel, Part I.) Fcp. 8vo. 5*s.*

King David. (The Tragedy of Israel, Part II.) Fcp. 8vo. 6*s.*

King Solomon. (The Tragedy of Israel, Part III.) Fcp. 8vo. 6*s.*

Ugone : A Tragedy. Fcp. 8vo. 6*s.*

A Garland from Greece ; Poems. Fcp. 8vo. 9*s.*

Stories of Wicklow ; Poems. Fcp. 8vo. 9*s.*

Victoria Regina et Imperatrix : a Jubilee Song from Ireland, 1887. 4to. 2*s.* 6*d.*

Mephistopheles in Broadcloth : a Satire. Fcp. 8vo. 4*s.*

The Life and Letters of Edmund J. Armstrong. Fcp. 8vo. 7*s.* 6*d.*

ARMSTRONG (E. J.)—**Works by.**

Poetical Works. Fcp. 8vo. 5*s.*

Essays and Sketches. Fcp. 8vo. 5*s.*

ARNOLD. — **The Light of the World ;** or, the Great Consummation. A Poem. By Sir EDWIN ARNOLD, K.C.I.E. Crown 8vo. 7*s.* 6*d.* net.

ARNOLD (Dr. T.)—**Works by.**

Introductory Lectures on Modern History. 8vo. 7*s.* 6*d.*

Sermons Preached mostly in the Chapel of Rugby School. 6 vols. crown 8vo. 30*s.* or separately, 5*s.* ea.

Miscellaneous Works. 8vo. 7*s.* 6*d.*

ASHLEY.—**English Economic History and Theory.** By W. J. ASHLEY, M.A. Professor of Political Economy in the University of Toronto. Part I.—The Middle Ages. 5*s.*

Atelier (The) du Lys ; or, an Art Student in the Reign of Terror. By the Author of 'Mademoiselle Mori.' Crown 8vo. 2*s.* 6*d.*

BY THE SAME AUTHOR.

Mademoiselle Mori : a Tale of Modern Rome. Crown 8vo. 2*s.* 6*d.*

That Child. Illustrated by GORDON BROWNE. Crown 8vo. 2*s.* 6*d.*

Atelier (The) du Lys—Works by the Author of—*continued.*

Under a Cloud. Crown 8vo. 2*s.* 6*d.*

The Fiddler of Lugau. With Illustrations by W. RALSTON. Crown 8vo. 2*s.* 6*d.*

A Child of the Revolution. With Illustrations by C. J. STANILAND. Crown 8vo. 2*s.* 6*d.*

Hester's Venture: a Novel. Crown 8vo. 2*s.* 6*d.*

In the Olden Time: a Tale of the Peasant War in Germany. Crown. 8vo. 2*s.* 6*d.*

BACON.—The Works and Life of.

Complete Works. Edited by R. L. ELLIS, J. SPEDDING, and D. D. HEATH. 7 vols. 8vo. £3. 13*s.* 6*d.*

Letters and Life, including all his Occasional Works. Edited by J. SPEDDING. 7 vols. 8vo. £4. 4*s.*

The Essays; with Annotations. By RICHARD WHATELY, D.D., 8vo. 10*s.* 6*d.*

The Essays; with Introduction, Notes, and Index. By E. A. ABBOTT, D.D. 2 vols. fcp. 8vo. price 6*s.* Text and Index only, without Introduction and Notes, in 1 vol. fcp. 8vo. 2*s.* 6*d.*

The BADMINTON LIBRARY, edited by the DUKE OF BEAUFORT, K.G. assisted by ALFRED E. T. WATSON.

Hunting. By the DUKE OF BEAU-FORT, K.G. and MOWBRAY MORRIS. With 53 Illus. by J. Sturgess, J. Charlton, and A. M. Biddulph. Crown 8vo. 10*s.* 6*d.*

Fishing. By H. CHOLMONDELEY-PENNELL.
Vol. I. Salmon, Trout, and Grayling. With 158 Illustrations. Cr. 8vo. 10*s.* 6*d.*
Vol. II. Pike and other Coarse Fish. With 132 Illustrations. Cr. 8vo. 10*s.* 6*d.*

Racing and Steeplechasing. By the EARL OF SUFFOLK AND BERKSHIRE, W. G. CRAVEN, &c. With 56 Illustrations by J. Sturgess. Cr. 8vo. 10*s.* 6*d.*

Shooting. By Lord WALSINGHAM and Sir RALPH PAYNE-GALLWEY, Bart.
Vol. I. Field and Covert. With 105 Illustrations. Cr. 8vo. 10*s.* 6*d.*
Vol. II. Moor and Marsh. With 65 Illustrations. Cr. 8vo. 10*s.* 6*d.*

The BADMINTON LIBRARY —*continued.*

Cycling. By VISCOUNT BURY, K.C.M.G. and G. LACY HILLIER. With 19 Plates and 70 Woodcuts by Viscount Bury, Joseph Pennell, &c. Cr. 8vo. 10*s.* 6*d.*

Athletics and Football. By MONTAGUE SHEARMAN. With 6 full-page Illustrations and 45 Woodcuts by Stanley Berkeley, and from Photographs by G. Mitchell. Cr. 8vo. 10*s.* 6*d.*

Boating. By W. B. WOODGATE With 10 full-page Illustrations and 39 Woodcuts in the Text. Cr. 8vo. 10*s.* 6*d.*

Cricket. By A. G. STEEL and the Hon. R. H. LYTTELTON. With 11 full-page Illustrations and 52 Woodcuts in the Text, by Lucien Davis. Cr. 8vo. 10*s.* 6*d.*

Driving. By the DUKE OF BEAUFORT. With 11 Plates and 54 Woodcuts by J. Sturgess and G. D. Giles. Cr. 8vo. 10*s.* 6*d.*

Fencing, Boxing, and Wrestling. By WALTER H. POLLOCK, F. C. GROVE, C. PREVOST, E. B. MICHELL, and WALTER ARMSTRONG. With 18 Plates and 24 Woodcuts. Crown 8vo. 10*s.* 6*d.*

Golf. By HORACE HUTCHINSON, the Rt. Hon. A. J. BALFOUR, M.P. ANDREW LANG, Sir W. G. SIMPSON, Bart. &c. With 19 Plates and 69 Woodcuts. Crown 8vo. 10*s.* 6*d.*

Tennis, Lawn Tennis, Rackets, and Fives. By J. M. and C. G. HEATHCOTE, E. O. PLEYDELL-BOUVERIE, and A. C. AINGER. With 12 Plates and 67 Woodcuts, &c. Crown 8vo. 10*s.* 6*d.*

BAGEHOT (Walter)—Works by.

Biographical Studies. 8vo. 12*s.*

Economic Studies. 8vo. 10*s.* 6*d.*

Literary Studies. 2 vols. 8vo. 28*s.*

The Postulates of English Political Economy. Cr. 8vo. 2*s.* 6*d.*

A Practical Plan for Assimilating the English and American Money as a Step towards a Universal Money. Cr. 8vo. 2*s.* 6*d.*

BAGWELL. — Ireland under the Tudors, with a Succinct Account of the Earlier History. By RICHARD BAGWELL, M.A. (3 vols.) Vols. I. and II. From the first invasion of the Northmen to the year 1578. 8vo. 32*s.* Vol. III. 1578–1603. 8vo. 18*s.*

A 2

BAIN (Alexander)—Works by.

Mental and Moral Science.
Crown 8vo. 10s. 6d.

Senses and the Intellect. 8vo. 15s.

Emotions and the Will. 8vo. 15s.

Logic, Deductive and Inductive.
PART I. *Deduction*, 4s. PART II. *Induction*, 6s. 6d.

Practical Essays. Cr. 8vo. 2s.

BAKER.—By the Western Sea :
a Summer Idyll. By JAMES BAKER,
F.R.G.S. Author of 'John Westacott.'
Cr. 8vo. 6s.

BAKER (Sir S. W.)—Works by.

Eight Years in Ceylon. With
6 Illustrations. Crown 8vo. 3s. 6d.

The Rifle and the Hound in
Ceylon. With 6 Illustrations. Crown
8vo. 3s. 6d.

BALL (The Rt. Hon. J. T.)—Works by.

The Reformed Church of Ireland
(1537-1889). 8vo. 7s. 6d.

Historical Review of the Legis-
lative Systems Operative in
Ireland, from the Invasion of Henry the
Second to the Union (1172-1800). 8vo. 6s.

BEACONSFIELD (The Earl of) —
Works by.

Novels and Tales. The Hugh-
enden Edition. With 2 Portraits and 11
Vignettes. 11 vols. Crown 8vo. 42s.

Endymion.	Henrietta Temple.
Lothair.	Contarini Fleming, &c.
Coningsby.	Alroy, Ixion, &c.
Tancred. Sybil.	The Young Duke, &c.
Venetia.	Vivian Grey.

Novels and Tales. Cheap Edition.
complete in 11 vols. Crown 8vo. 1s.
each, boards ; 1s. 6d. each, cloth.

BECKER (Professor)—Works by.

Gallus; or, Roman Scenes in the
Time of Augustus. Post 8vo. 7s. 6d.

Charicles; or, Illustrations of the
Private Life of the Ancient Greeks.
Post 8vo. 7s. 6d.

BELL (Mrs. Hugh).—Works by.

Will o' the Wisp: a Story.
Illustrated by E. L. SHUTE. Crown 8vo.
3s. 6d.

Chamber Comedies: a Collection
of Plays and Monologues for the Drawing
Room. Crown 8vo. 6s.

BLAKE.—Tables for the Conversion of 5 per Cent. Interest
from $\frac{1}{16}$ to 7 per Cent. By J.
BLAKE, of the London Joint Stock Bank,
Limited. 8vo. 12s. 6d.

Book (The) of Wedding Days.
Arranged on the Plan of a Birthday Book.
With 96 Illustrated Borders, Frontispiece,
and Title-page by WALTER CRANE ; and
Quotations for each Day. Compiled and
Arranged by K. E. J. REID, MAY ROSS,
and MABEL BAMFIELD. 4to. 21s.

BRASSEY (Lady)—Works by.

A Voyage in the 'Sunbeam,' our
Home on the Ocean for
Eleven Months.

Library Edition. With 8 Maps and
Charts, and 118 Illustrations, 8vo. 21s.

Cabinet Edition. With Map and 66
Illustrations, crown 8vo. 7s. 6d.

School Edition. With 37 Illustrations,
fcp. 2s. cloth, or 3s. white parchment.

Popular Edition. With 60 Illustrations,
4to. 6d. sewed, 1s. cloth.

Sunshine and Storm in the East.

Library Edition. With 2 Maps and
114 Illustrations, 8vo. 21s.

Cabinet Edition. With 2 Maps and
114 Illustrations, crown 8vo. 7s. 6d.

Popular Edition. With 103 Illustrations, 4to. 6d. sewed, 1s. cloth.

In the Trades, the Tropics, and
the 'Roaring Forties.'

Cabinet Edition. With Map and 220
Illustrations, crown 8vo. 7s. 6d.

Popular Edition. With 183 Illustrations, 4to. 6d. sewed, 1s. cloth.

BRASSEY (Lady) — Works by — *continued.*

The Last Voyage to India and Australia in the 'Sunbeam.' With Charts and Maps, and 40 Illustrations in Monotone (20 full-page), and nearly 200 Illustrations in the Text from Drawings by R. T. PRITCHETT. 8vo. 21*s.*

Three Voyages in the 'Sunbeam.' Popular Edition. With 346 Illustrations, 4to. 2*s.* 6*d.*

BRAY.—The Philosophy of Necessity; or, Law in Mind as in Matter. By CHARLES BRAY. Crown 8vo. 5*s.*

BRIGHT.--A History of England. By the Rev. J. FRANCK BRIGHT, D.D. Master of University College, Oxford. 4 vols. crown 8vo.

Period I.—Mediæval Monarchy : The Departure of the Romans to Richard III. From A.D. 449 to 1485. 4*s.* 6*d.*

Period II.—Personal Monarchy : Henry VII. to James II. From 1485 to 1688. 5*s.*

Period III. – Constitutional Monarchy : William and Mary to William IV. From 1689 to 1837. 7*s.* 6*d.*

Period IV.—The Growth of Democracy : Victoria. From 1837 to 1880. 6*s.*

BRYDEN. — Kloof and Karroo: Sport, Legend, and Natural History in Cape Colony. By H. A. BRYDEN. With 17 Illustrations. 8vo. 10*s.* 6*d.*

BUCKLE. — History of Civilisation in England and France, Spain and Scotland. By HENRY THOMAS BUCKLE. 3 vols. cr. 8vo. 24*s.*

BUCKTON (Mrs. C. M.)—Works by.

Food and Home Cookery. With 11 Woodcuts. Crown 8vo. 2*s.* 6*d.*

Health in the House. With 41 Woodcuts and Diagrams. Crown 8vo. 2*s.*

BULL (Thomas)—Works by.

Hints to Mothers on the Management of their Health during the Period of Pregnancy. Fcp. 8vo. 1*s.* 6*d.*

The Maternal Management of Children in Health and Disease. Fcp. 8vo. 1*s.* 6*d.*

BUTLER (Samuel)—Works by.

Op. 1. Erewhon. Cr. 8vo. 5*s.*

Op. 2. The Fair Haven. A Work in Defence of the Miraculous Element in our Lord's Ministry. Cr. 8vo. 7*s.* 6*d.*

Op. 3. Life and Habit. An Essay after a Completer View of Evolution. Cr. 8vo. 7*s.* 6*d.*

Op. 4. Evolution, Old and New. Cr. 8vo. 10*s.* 6*d.*

Op. 5. Unconscious Memory. Cr. 8vo. 7*s.* 6*d.*

Op. 6. Alps and Sanctuaries of Piedmont and the Canton Ticino. Illustrated. Pott 4to. 10*s.* 6*d.*

Op. 7. Selections from Ops. 1–6. With Remarks on Mr. G. J. ROMANES' 'Mental Evolution in Animals.' Cr. 8vo. 7*s.* 6*d.*

Op. 8. Luck, or Cunning, as the Main Means of Organic Modification? Cr. 8vo. 7*s.* 6*d.*

Op. 9. Ex Voto. An Account of the Sacro Monte or New Jerusalem at Varallo-Sesia. 10*s.* 6*d.*

Holbein's 'La Danse.' A Note on a Drawing called 'La Danse.' 3*s.*

CARLYLE. — Thomas Carlyle: a History of his Life. By J. A. FROUDE. 1795-1835, 2 vols. crown 8vo. 7*s.* 1834-1881, 2 vols. crown 8vo. 7*s.*

CASE. — Physical Realism: being an Analytical Philosophy from the Physical Objects of Science to the Physical Data of Sense. By THOMAS CASE, M.A. Fellow and Senior Tutor C.C.C. 8vo. 15*s.*

CHETWYND. — Racing Reminiscences and Experiences of the Turf. By Sir GEORGE CHETWYND, Bart. 2 vols. 8vo. 21*s.*

CHILD. — Church and State under the Tudors. By GILBERT W. CHILD, M.A. Exeter College, Oxford. 8vo. 15*s.*

CHISHOLM.—Handbook of Commercial Geography. By G. G. CHISHOLM, B.Sc. With 29 Maps. 8vo. 16*s.*

CHURCH.—Sir Richard Church, C.B. G.C.H. Commander-in-Chief of the Greeks in the War of Independence: a Memoir. By STANLEY LANE-POOLE, Author of 'The Life of Viscount Stratford de Redcliffe.' With 2 Plans. 8vo. 5*s.*

CLARK-KENNEDY.—Pictures in Rhyme. By ARTHUR CLARK-KENNEDY. With Illustrations by MAURICE GREIFFENHAGEN. Cr. 8vo.

CLIVE.—Poems. By V. (Mrs. ARCHER CLIVE), Author of 'Paul Ferroll.' Including the IX. Poems. New Edition. Fcp. 8vo. 6*s.*

CLODD.—The Story of Creation : a Plain Account of Evolution. By EDWARD CLODD. With 77 Illustrations. Crown 8vo. 3*s.* 6*d.*

CLUTTERBUCK.—The Skipper in Arctic Seas. By W. J. CLUTTERBUCK, one of the Authors of 'Three in Norway.' With 39 Illustrations. Cr. 8vo. 10*s.* 6*d.*

COLENSO.—The Pentateuch and Book of Joshua Critically Examined. By J. W. COLENSO, D.D. late Bishop of Natal. Crown 8vo. 6*s.*

COLMORE.—A Living Epitaph. By G. COLMORE, Author of 'A Conspiracy of Silence' &c. Crown 8vo. 6*s.*

COMYN. — Atherstone Priory : a Tale. By L. N. COMYN. Cr. 8vo. 2*s.* 6*d.*

CONINGTON (John)—Works by.

The Æneid of Virgil. Translated into English Verse. Crown 8vo. 6*s.*

The Poems of Virgil. Translated into English Prose. Crown 8vo. 6*s.*

COX. — A General History of Greece, from the Earliest Period to the Death of Alexander the Great; with a sketch of the subsequent History to the Present Time. By the Rev. Sir G. W. COX, Bart. M.A. With 11 Maps and Plans. Crown 8vo. 7*s.* 6*d.*

CRAKE. — Historical Tales. By A. D. CRAKE, B.A. Author of 'History of the Church under the Roman Empire,' &c. &c. Crown 8vo. 5 vols. 3*s.* 6*d.* each. Sold separately.

Edwy the Fair ; or, The First Chronicle of Æscendune.

Alfgar the Dane ; or, The Second Chronicle of Æscendune.

The Rival Heirs : being the Third and Last Chronicle of Æscendune.

The House of Walderne. A Tale of the Cloister and the Forest in the Days of the Barons' Wars.

Brian Fitz-Count. A Story of Wallingford Castle and Dorchester Abbey.

CRAKE.—History of the Church under the Roman Empire, A.D. 30–476. By the Rev. A. D. CRAKE, B.A. late Vicar of Cholsey, Berks. Crown 8vo. 7*s.* 6*d.*

CREIGHTON. — History of the Papacy During the Reformation. By MANDELL CREIGHTON, D.D. LL.D. Bishop of Peterborough. 8vo. Vols. I. and II. 1378–1464, 32*s.* ; Vols. III. and IV. 1464–1518, 24*s.*

CRUMP (A.)—Works by.

A Short Enquiry into the Formation of Political Opinion, from the Reign of the Great Families to the Advent of Democracy. 8vo. 7*s.* 6*d.*

An Investigation into the Causes of the Great Fall in Prices which took place coincidently with the Demonetisation of Silver by Germany. 8vo. 6*s.*

CURZON.—Russia in Central Asia in 1889 and the Anglo-Russian Question. By the Hon. GEORGE N. CURZON, M.P. 8vo. 21*s.*

DANTE.—La Commedia di Dante. A New Text, carefully Revised with the aid of the most recent Editions and Collations. Small 8vo. 6*s.*
*** Fifty Copies (of which Forty-five are for Sale) have been printed on Japanese paper, £1. 1*s.* net.

DAVIDSON (W. L.)—Works by.

The Logic of Definition Explained and Applied. Cr. 8vo. 6s.

Leading and Important English Words Explained and Exemplified. Fcp. 8vo. 3s. 6d.

DELAND (Mrs.)—Works by.

John Ward, Preacher: a Story. Crown 8vo. 2s. boards, 2s. 6d. cloth.

Sidney: a Novel. Crown 8vo. 6s.

The Old Garden, and other Verses. Fcp. 8vo. 5s.

Florida Days. With 12 Full-page Plates (2 Etched and 4 in Colours), and about 50 Illustrations in the Text, by Louis K. Harlow. 8vo. 21s.

DE LA SAUSSAYE.—A Manual of the Science of Religion. By Professor Chantepie de la Saussaye. Translated by Mrs. Colyer Fergusson (née Max Müller). Revised by the Author.

DE REDCLIFFE.—The Life of the Right Hon. Stratford Canning: Viscount Stratford De Redcliffe. By Stanley Lane-Poole.

Cabinet Edition, abridged, with 3 Portraits, 1 vol. crown 8vo. 7s. 6d.

DE SALIS (Mrs.)—Works by.

Savouries à la Mode. Fcp. 8vo. 1s. 6d. boards.

Entrées à la Mode. Fcp. 8vo. 1s. 6d. boards.

Soups and Dressed Fish à la Mode. Fcp. 8vo. 1s. 6d. boards.

Oysters à la Mode. Fcp. 8vo. 1s. 6d. boards.

Sweets and Supper Dishes à la Mode. Fcp. 8vo. 1s. 6d. boards.

Dressed Vegetables à la Mode. Fcp. 8vo. 1s. 6d. boards.

Dressed Game and Poultry à la Mode. Fcp. 8vo. 1s. 6d. boards.

Puddings and Pastry à la Mode. Fcp. 8vo. 1s. 6d. boards.

DE SALIS (Mrs.)—Works by—cont.

Cakes and Confections à la Mode. Fcp. 8vo. 1s. 6d. boards.

Tempting Dishes for Small Incomes. Fcp. 8vo. 1s. 6d.

Wrinkles and Notions for every Household. Crown 8vo. 2s. 6d.

DE TOCQUEVILLE.—Democracy in America. By Alexis de Tocqueville. Translated by Henry Reeve, C.B. 2 vols. crown 8vo. 16s.

DOWELL.—A History of Taxation and Taxes in England from the Earliest Times to the Year 1885. By Stephen Dowell. (4 vols. 8vo.) Vols. I. and II. The History of Taxation, 21s. Vols. III. and IV. The History of Taxes, 21s.

DOYLE (A. Conan)—Works by.

Micah Clarke: his Statement as made to his three Grandchildren, Joseph, Gervas, and Reuben, during the hard Winter of 1734. With Frontispiece and Vignette. Crown 8vo. 3s. 6d.

The Captain of the Polestar; and other Tales. Crown 8vo. 6s.

Dublin University Press Series (The): a Series of Works undertaken by the Provost and Senior Fellows of Trinity College, Dublin.

Abbott's (T. K.) Codex Rescriptus Dublinensis of St. Matthew. 4to. 21s.

———— Evangeliorum Versio Antehieronymiana ex Codice Usseriano (Dublinensi). 2 vols. crown 8vo. 21s.

Allman's (G. J.) Greek Geometry from Thales to Euclid. 8vo. 10s. 6d.

Burnside (W. S.) and Panton's (A. W.) Theory of Equations. 8vo. 12s. 6d.

Casey's (John) Sequel to Euclid's Elements. Crown 8vo. 3s. 6d.

———— Analytical Geometry of the Conic Sections. Crown 8vo. 7s. 6d.

Davies' (J. F.) Eumenides of Æschylus. With Metrical English Translation. 8vo. 7s.

Dublin Translations into Greek and Latin Verse. Edited by R. Y. Tyrrell. 8vo. 6s.

[*Continued on next page.*

Dublin University Press Series (The)—*continued.*

Graves' (R. P.) Life of Sir William Hamilton. 3 vols. 15*s.* each.

Griffin (R. W.) on Parabola, Ellipse, and Hyperbola. Crown 8vo. 6*s.*

Hobart's (W. K.) Medical Language of St. Luke. 8vo. 16*s.*

Leslie's (T. E. Cliffe) Essays in Political Economy. 8vo. 10*s.* 6*d.*

Macalister's (A.) Zoology and Morphology of Vertebrata. 8vo. 10*s.* 6*d.*

MacCullagh's (James) Mathematical and other Tracts. 8vo. 15*s.*

Maguire's (T.) Parmenides of Plato, Text with Introduction, Analysis, &c. 8vo. 7*s.* 6*d.*

Monck's (W. H. S.) Introduction to Logic. Crown 8vo. 5*s.*

Roberts' (R. A.) Examples in the Analytic 5*s.*

Southey's (R.) Correspondence with Caroline Bowles. Edited by E. Dowden. 8vo. 14*s.*

Stubbs' (J. W.) History of the University of Dublin, from its Foundation to the End of the Eighteenth Century. 8vo. 12*s.* 6*d.*

Thornhill's (W. J.) The Æneid of Virgil, freely translated into English Blank Verse. Crown 8vo. 7*s.* 6*d.*

Tyrrell's (R. Y.) Cicero's Correspondence. Vols. I. II. and III. 8vo. each 12*s.*

————— The Acharnians of Aristophanes, translated into English Verse. Crown 8vo. 1*s.*

Webb's (T. E.) Goethe's Faust, Translation and Notes. 8vo. 12*s.* 6*d.*

————— The Veil of Isis : a Series of Essays on Idealism. 8vo. 10*s.* 6*d.*

Wilkins' (G.) The Growth of the Homeric Poems. 8vo. 6*s.*

Epochs of Modern History.
Edited by C. Colbeck, M.A. 19 vols. fcp. 8vo. with Maps, 2*s.* 6*d.* each.

Church's (Very Rev. R. W.) The Beginning of the Middle Ages. With 3 Maps.

Johnson's (Rev. A. H.) The Normans in Europe. With 3 Maps.

Cox's (Rev. Sir G. W.) The Crusades. With a Map.

Stubbs's (Right Rev. W.) The Early Plantagenets. With 2 Maps.

Warburton's (Rev. W.) Edward the Third. With 3 Maps and 3 Genealogical Tables.

Epochs of Modern History—*continued.*

Gairdner's (J.) The Houses of Lancaster and York ; with the Conquest and Loss of France. With 5 Maps.

Moberly's (Rev. C. E.) The Early Tudors.

Seebohm's (F.) The Era of the Protestant Revolution. With 4 Maps and 12 Diagrams.

Creighton's (Rev. M.) The Age of Elizabeth. With 5 Maps and 4 Genealogical Tables.

Gardiner's (S. R.) The First Two Stuarts and the Puritan Revolution (1603-1660). With 4 Maps.

Gardiner's (S. R.) The Thirty Years' War (1618-1648). With a Map.

Airy's (O.) The English Restoration and Louis XIV. (1648-1678).

Hale's (Rev. E.) The Fall of the Stuarts ; and Western Europe (1678-1697). With 11 Maps and Plans.

Morris's (E. E.) The Age of Anne. With 7 Maps and Plans.

Morris's (E. E.) The Early Hanoverians. With 9 Maps and Plans.

Longman's (F. W.) Frederick the Great and the Seven Years' War. With 2 Maps.

Ludlow's (J. M.) The War of American Independence (1775-1783). With 4 Maps.

Gardiner's (Mrs. S. R.) The French Revolution (1789-1795). With 7 Maps.

McCarthy's (Justin) The Epoch of Reform (1830-1850).

Epochs of Church History.
Edited by Mandell Creighton, D.D. Bishop of Peterborough. Fcp. 8vo. 2*s.* 6*d.* each.

Tucker's (Rev. H. W.) The English Church in other Lands.

Perry's (Rev. G. G.) The History of the Reformation in England.

Brodrick's (Hon. G. C.) A History of the University of Oxford.

Mullinger's (J. B.) A History of the University of Cambridge.

Plummer's (A.) The Church of the Early Fathers.

Carr's (Rev. A.) The Church and the Roman Empire.

Wakeman's (H. O.) The Church and the Puritans (1570-1660).

Overton's (Rev. J. H.) The Evangelical Revival in the Eighteenth Century.

Tozer's (Rev. H. F.) The Church and the Eastern Empire.

Epochs of Church History—
continued.

Stephens's (Rev. W. R.W.) Hildebrand and his Times.

Hunt's (Rev. W.) The English Church in the Middle Ages.

Balzani's (U.) The Popes and the Hohenstaufen.

Gwatkin's (H. M.) The Arian Controversy.

Ward's (A. W.) The Counter-Reformation.

Poole's (R. L.) Wycliffe and Early Movements of Reform.

Epochs of Ancient History.
Edited by the Rev. Sir G. W. Cox, Bart. M.A. and by C. Sankey, M.A. 10 volumes, fcp. 8vo. with Maps, 2*s.* 6*d.* each.

Beesly's (A. H.) The Gracchi, Marius, and Sulla. With 2 Maps.

Capes's (Rev. W. W.) The Early Roman Empire. From the Assassination of Julius Cæsar to the Assassination of Domitian. With 2 Maps.

———————— The Roman Empire of the Second Century, or the Age of the Antonines. With 2 Maps.

Cox's (Rev. Sir G. W.) The Athenian Empire from the Flight of Xerxes to the Fall of Athens. With 5 Maps.

———————— The Greeks and the Persians. With 4 Maps.

Curteis's (A. M.) The Rise of the Macedonian Empire. With 8 Maps.

Ihne's (W.) Rome to its Capture by the Gauls. With a Map.

Merivale's (Very Rev. C.) The Roman Triumvirates. With a Map.

Sankey's (C.) The Spartan and Theban Supremacies. With 5 Maps.

Smith's (R. B.) Rome and Carthage, the Punic Wars. With 9 Maps and Plans.

Epochs of American History.
Edited by Dr. Albert Bushnell Hart, Assistant Professor of History in Harvard College.

Thwaites's (R. G.) The Colonies (1492–1763). Fcp. 8vo. 3*s.* 6*d.* [*Ready.*

Hart's (A. B.) Formation of the Union (1763–1829). Fcp. 8vo. [*In preparation.*

Wilson's (W.) Division and Re-union (1829–1889). Fcp. 8vo. [*In preparation.*

Epochs of English History.
Complete in One Volume, with 27 Tables and Pedigrees, and 23 Maps. Fcp. 8vo. 5*s.*

*** For details of Parts *see* Longmans & Co.'s Catalogue of School Books.

EWALD (Heinrich)—Works by.

The Antiquities of Israel. Translated from the German by H. S. Solly, M.A. 8vo. 12*s.* 6*d.*

The History of Israel. Translated from the German. 8 vols. 8vo. Vols. I. and II. 24*s.* Vols. III. and IV. 21*s.* Vol. V. 18*s.* Vol. VI. 16*s.* Vol. VII. 21*s.* Vol. VIII. with Index to the Complete Work. 18*s.*

FARNELL.—The Greek Lyric Poets. Edited, with Introductions and Notes, by G. S. Farnell, M.A. 8vo.

FARRAR.—Language and Languages. A Revised Edition of *Chapters on Language and Families of Speech.* By F. W. Farrar, D.D. Crown 8vo. 6*s.*

FIRTH.—Nation Making: a Story of New Zealand Savageism and Civilisation. By J. C. Firth, Author of 'Luck' and 'Our Kin across the Sea.' Crown 8vo. 6*s.*

FITZWYGRAM. — Horses and Stables. By Major-General Sir F. Fitzwygram, Bart. With 19 pages of Illustrations. 8vo. 5*s.*

FORD.—The Theory and Practice of Archery. By the late Horace Ford. New Edition, thoroughly Revised and Re-written by W. Butt, M.A. With a Preface by C. J. Longman, M.A, F.S.A. 8vo. 14*s.*

FOUARD.—The Christ the Son of God: a Life of our Lord and Saviour Jesus Christ. By the Abbé Constant Fouard. Translated from the Fifth Edition, with the Author's sanction, by George F. X. Griffith. With an Introduction by Cardinal Manning. 2 vols. crown 8vo. 14*s.*

FOX.—The Early History of Charles James Fox. By the Right Hon. Sir G. O. Trevelyan, Bart. Library Edition, 8vo. 18*s.* Cabinet Edition, cr. 8vo. 6*s.*

FRANCIS—A Book on Angling; or, Treatise on the Art of Fishing in every branch; including full Illustrated List of Salmon Flies. By Francis Francis. Post 8vo. Portrait and Plates, 15*s.*

FREEMAN.—The Historical Geography of Europe. By E. A. Freeman. With 65 Maps. 2 vols. 8vo. 31*s.* 6*d.*

FROUDE (James A.)—Works by.

The History of England, from the Fall of Wolsey to the Defeat of the Spanish Armada. 12 vols. crown 8vo. £2. 2s.

Short Studies on Great Subjects. Cabinet Edition, 4 vols. crown 8vo. 24s. Cheap Edition, 4 vols. crown 8vo. 3s. 6d. each.

Cæsar : a Sketch. Crown 8vo. 3s. 6d.

The English in Ireland in the Eighteenth Century. 3 vols. crown 8vo. 18s.

Oceana; or, England and Her Colonies. With 9 Illustrations. Crown 8vo. 2s. boards, 2s. 6d. cloth.

The English in the West Indies; or, the Bow of Ulysses. With 9 Illustrations. Crown 8vo. 2s. boards, 2s. 6d. cloth.

The Two Chiefs of Dunboy; an Irish Romance of the Last Century. Crown 8vo. 6s.

Thomas Carlyle, a History of his Life. 1795 to 1835. 2 vols. crown 8vo. 7s. 1834 to 1881. 2 vols. crown 8vo. 7s.

GALLWEY.—Letters to Young Shooters. (First Series.) On the Choice and Use of a Gun. By Sir RALPH PAYNE-GALLWEY, Bart. With Illustrations. Crown 8vo. 7s. 6d.

GARDINER (Samuel Rawson)— Works by.

History of England, from the Accession of James I. to the Outbreak of the Civil War, 1603-1642. 10 vols. crown 8vo. price 6s. each.

A History of the Great Civil War, 1642-1649. (3 vols.) Vol. I. 1642-1644. With 24 Maps. 8vo. 21s. (out of print). Vol. II. 1644-1647. With 21 Maps. 8vo. 24s.

The Student's History of England. Illustrated under the superintendence of Mr. ST. JOHN HOPE, Secretary to the Society of Antiquaries. Vol. I. B.C. 55—A.D. 1509, with 173 Illustrations, crown 8vo. 4s. Vol. II. 1509-1689, with 96 Illustrations. Crown 8vo. 4s.

The work will be published in Three Volumes, and also in One Volume complete.

GIBERNE—Works by.

Ralph Hardcastle's Will. By AGNES GIBERNE. With Frontispiece. Crown 8vo. 5s.

Nigel Browning. Crown 8vo. 5s.

GOETHE.—Faust. A New Translation chiefly in Blank Verse ; with Introduction and Notes. By JAMES ADEY BIRDS. Crown 8vo. 6s.

Faust. The Second Part. A New Translation in Verse. By JAMES ADEY BIRDS. Crown 8vo. 6s.

GREEN.—The Works of Thomas Hill Green. Edited by R. L. NETTLESHIP (3 vols.) Vols. I. and II.— Philosophical Works. 8vo. 16s. each. Vol. III.—Miscellanies. With Index to the three Volumes and Memoir. 8vo. 21s.

The Witness of God and Faith : Two Lay Sermons. By T. H. GREEN. Fcp. 8vo. 2s.

GREVILLE. — A Journal of the Reigns of King George IV. King William IV. and Queen Victoria. By C. C. F. GREVILLE. Edited by H. REEVE. 8 vols. Cr. 8vo. 6s. ea.

GREY.—Last Words to Girls. On Life in School and after School. By Mrs. WILLIAM GREY. Cr 8vo. 3s. 6d.

GWILT. — An Encyclopædia of Architecture. By JOSEPH GWILT, F.S.A. Illustrated with more than 1,700 Engravings on Wood. 8vo. 52s. 6d.

HAGGARD.—Life and its Author : an Essay in Verse. By ELLA HAGGARD. With a Memoir by H. RIDER HAGGARD, and Portrait. Fcp. 8vo. 3s. 6d.

HAGGARD (H. Rider)—Works by.

She. With 32 Illustrations by M. GREIFFENHAGEN and C. H. M. KERR. Crown 8vo. 3s. 6d.

Allan Quatermain. With 31 Illustrations by C. H. M. KERR. Crown 8vo. 3s. 6d.

Maiwa's Revenge ; or, the War of the Little Hand. Crown 8vo 2s. boards ; 2s. 6d. cloth.

Colonel Quaritch, V.C. A Novel. Crown 8vo. 3s. 6d.

HAGGARD (H. Rider)—Works by—
continued.

Cleopatra : being an Account of the Fall and Vengeance of Harmachis, the Royal Egyptian. With 29 Full-page Illustrations by M. Greiffenhagen and R. Caton Woodville. Crown 8vo. 3s. 6d.

Beatrice. A Novel. Cr. 8vo. 6s.

HAGGARD and LANG.—The World's Desire.
By H. RIDER HAGGARD and ANDREW LANG. Crown 8vo. 6s.

HARRISON.—Myths of the Odyssey in Art and Literature.
Illustrated with Outline Drawings. By JANE E. HARRISON. 8vo. 18s.

HARRISON. — The Contemporary History of the French Revolution,
compiled from the 'Annual Register.' By F. BAYFORD HARRISON. Crown 8vo. 3s. 6d.

HARTE (Bret)—Works by.
In the Carquinez Woods. Fcp. 8vo. 1s. boards; 1s. 6d. cloth.

On the Frontier. 16mo. 1s.

By Shore and Sedge. 16mo. 1s.

HARTWIG (Dr.)—Works by.
The Sea and its Living Wonders. With 12 Plates and 303 Woodcuts. 8vo. 10s. 6d.

The Tropical World. With 8 Plates, and 172 Woodcuts. 8vo. 10s. 6d.

The Polar World. With 3 Maps, 8 Plates, and 85 Woodcuts. 8vo. 10s. 6d.

The Subterranean World. With 3 Maps and 80 Woodcuts. 8vo. 10s. 6d.

The Aerial World. With Map, 8 Plates, and 60 Woodcuts. 8vo. 10s. 6d.

The following books are extracted from the foregoing works by Dr. HARTWIG :—

Heroes of the Arctic Regions. With 19 Illustrations. Crown 8vo. 2s.

Wonders of the Tropical Forests. With 40 Illustrations. Crown 8vo. 2s.

Workers Under the Ground. or, Mines and Mining. With 29 Illustrations. Crown 8vo. 2s.

Marvels Over Our Heads. With 29 Illustrations. Crown 8vo. 2s.

Marvels Under Our Feet. With 22 Illustrations. Crown 8vo. 2s.

HARTWIG (Dr.)—Works by—*cont.*
Dwellers in the Arctic Regions. With 29 Illustrations. Crown 8vo. 2s. 6d.

Winged Life in the Tropics. With 55 Illustrations. Crown 8vo. 2s. 6d.

Volcanoes and Earthquakes. With 30 Illustrations. Crown 8vo. 2s. 6d.

Wild Animals of the Tropics. With 66 Illustrations. Crown 8vo. 3s. 6d.

Sea Monsters and Sea Birds. With 75 Illustrations. Crown 8vo. 2s. 6d.

Denizens of the Deep. With 117 Illustrations. Crown 8vo. 2s. 6d.

HAVELOCK. — Memoirs of Sir Henry Havelock, K.C.B.
By JOHN CLARK MARSHMAN. Cr. 8vo. 3s. 6d.

HEARN (W. Edward)—Works by.
The Government of England; its Structure and its Development. 8vo. 16s.

The Aryan Household : its Structure and its Development. An Introduction to Comparative Jurisprudence. 8vo. 16s.

HISTORIC TOWNS.
Edited by E. A. FREEMAN, D.C.L. and Rev. WILLIAM HUNT, M.A. With Maps and Plans. Crown 8vo. 3s. 6d. each.

Bristol. By Rev. W. HUNT.

Carlisle. By Rev. MANDELL CREIGHTON.

Cinque Ports. By MONTAGU BURROWS.

Colchester. By Rev. E. L. CUTTS.

Exeter. By E. A. FREEMAN.

London. By Rev. W. J. LOFTIE.

Oxford. By Rev. C. W. BOASE.

Winchester. By Rev. G. W. KITCHIN, D.D.

New York. By THEODORE ROOSEVELT.

Boston (U.S.) By HENRY CABOT LODGE. [*In the press.*

York. By Rev. JAMES RAINE. [*In preparation.*

HODGSON (Shadworth H.)—Works by.
Time and Space : a Metaphysical Essay. 8vo. 16s.

The Theory of Practice : an Ethical Enquiry. 2 vols. 8vo. 24s.

The Philosophy of Reflection : 2 vols. 8vo. 21s.

[*Continued on next page.*

HODGSON (Shadworth H.)—Works by—*continued.*

Outcast Essays and Verse Translations. Essays: The Genius of De Quincey—De Quincey as Political Economist—The Supernatural in English Poetry; with Note on the True Symbol of Christian Union — English Verse. Verse Translations: Nineteen Passages from Lucretius, Horace, Homer, &c. Crown 8vo. 8*s*. 6*d*.

HOWITT.—Visits to Remarkable Places, Old Halls, Battle-Fields, Scenes illustrative of Striking Passages in English History and Poetry. By WILLIAM HOWITT. 80 Illustrations. Cr. 8vo. 3*s*. 6*d*.

HULLAH (John)—Works by.

Course of Lectures on the History of Modern Music. 8vo. 8*s*. 6*d*.

Course of Lectures on the Transition Period of Musical History. 8vo. 10*s*. 6*d*.

HUME.—The Philosophical Works of David Hume. Edited by T. H. GREEN and T. H. GROSE. 4 vols. 8vo. 56*s*. Or separately, Essays, 2 vols. 28*s*. Treatise of Human Nature. 2 vols. 28*s*.

HUTCHINSON (Horace)—Works by.

Cricketing Saws and Stories. By HORACE HUTCHINSON. With rectilinear Illustrations by the Author. 16mo. 1*s*.

Some Great Golf Links. Edited by HORACE HUTCHINSON. With Illustrations.
This book is mainly a reprint of articles that have recently appeared in the *Saturday Review.*

HUTH.—The Marriage of Near Kin, considered with respect to the Law of Nations, the Result of Experience, and the Teachings of Biology. By ALFRED H. HUTH. Royal 8vo. 21*s*.

INGELOW (Jean)—Works by.

Poetical Works. Vols. I. and II. Fcp. 8vo. 12*s*. Vol. III. Fcp. 8vo. 5*s*.

Lyrical and Other Poems. Selected from the Writings of JEAN INGELOW. Fcp. 8vo. 2*s*. 6*d*. cloth plain; 3*s*. cloth gilt.

Very Young and Quite Another Story: Two Stories. Crown 8vo. 6*s*.

JAMES.—The Long White Mountain; or, a Journey in Manchuria, with an Account of the History, Administration, and Religion of that Province. By H. E. JAMES. With Illustrations. 8vo. 24*s*.

JAMESON (Mrs.)—Works by.

Legends of the Saints and Martyrs. With 19 Etchings and 187 Woodcuts. 2 vols. 8vo. 20*s*. *net.*

Legends of the Madonna, the Virgin Mary as represented in Sacred and Legendary Art. With 27 Etchings and 165 Woodcuts. 1 vol. 8vo. 10*s*. *net.*

Legends of the Monastic Orders. With 11 Etchings and 88 Woodcuts. 1 vol. 8vo. 10*s*. *net.*

History of Our Lord, His Types and Precursors. Completed by Lady EASTLAKE. With 31 Etchings and 281 Woodcuts. 2 vols. 8vo. 20*s*. *net.*

JEFFERIES.—Field and Hedgerow: last Essays of RICHARD JEFFERIES. Crown 8vo. 3*s*. 6*d*.

JENNINGS.—Ecclesia Anglicana. A History of the Church of Christ in England, from the Earliest to the Present Times. By the Rev. ARTHUR CHARLES JENNINGS, M.A. Crown 8vo. 7*s*. 6*d*.

JESSOP (G. H.)—Works by.

Judge Lynch: a Tale of the California Vineyards. Crown 8vo. 6*s*.

Gerald Ffrench's Friends. Cr. 8vo. 6*s*. A collection of Irish-American character stories.

JOHNSON. — The Patentee's Manual; a Treatise on the Law and Practice of Letters Patent. By J. JOHNSON and J. H. JOHNSON. 8vo. 10*s*. 6*d*.

JORDAN (William Leighton) — The Standard of Value. By WILLIAM LEIGHTON JORDAN. 8vo. 6*s*.

JUSTINIAN. — The Institutes of Justinian; Latin Text, chiefly that of Huschke, with English Introduction. Translation, Notes, and Summary. By THOMAS C. SANDARS, M.A. 8vo. 18*s*.

KALISCH (M. M.)—Works by.

Bible Studies. Part I. The Prophecies of Balaam. 8vo. 10*s*. 6*d*. Part II. The Book of Jonah. 8vo. 10*s*. 6*d*.

KALISCH (M. M.)—Works by—*contd.*

Commentary on the Old Testament; with a New Translation. Vol. I. Genesis, 8vo. 18*s.* or adapted for the General Reader, 12*s.* Vol. II. Exodus, 15*s.* or adapted for the General Reader, 12*s.* Vol. III. Leviticus, Part I. 15*s.* or adapted for the General Reader, 8*s.* Vol. IV. Leviticus, Part II. 15*s.* or adapted for the General Reader, 8*s.*

Hebrew Grammar. With Exercises. Part I. 8vo. 12*s.* 6*d.* Key, 5*s.* Part II. 12*s.* 6*d.*

KANT (Immanuel)—Works by.

Critique of Practical Reason, and other Works on the Theory of Ethics. Translated by T. K. Abbott, B.D. With Memoir. 8vo. 12*s.* 6*d.*

Introduction to Logic, and his Essay on the Mistaken Subtilty of the Four Figures. Translated by T. K. Abbott. Notes by S. T. Coleridge. 8vo. 6*s.*

KENDALL (May)—Works by.

From a Garret. Crown 8vo. 6*s.*

Dreams to Sell; Poems. Fcp. 8vo. 6*s.*

'Such is Life': a Novel. Crown 8vo. 6*s.*

KILLICK. — **Handbook to Mill's System of Logic.** By the Rev. A. H. KILLICK, M.A. Crown 8vo. 3*s.* 6*d.*

KNIGHT. — **The Cruise of the 'Alerte':** the Narrative of a Search for Treasure on the Desert Island of Trinidad. By E. F. KNIGHT, Author of 'The Cruise of the "Falcon."' With 2 Maps and 23 Illustrations. Crown 8vo. 10*s.* 6*d.*

LADD (George T.)—Works by.

Elements of Physiological Psychology. 8vo. 21*s.*

Outlines of Physiological Psychology. A Text-Book of Mental Science for Academies and Colleges. 8vo. 12*s.*

LANG (Andrew)—Works by.

Custom and Myth: Studies of Early Usage and Belief. With 15 Illustrations. Crown 8vo. 7*s.* 6*d.*

Books and Bookmen. With 2 Coloured Plates and 17 Illustrations. Cr. 8vo. 6*s.* 6*d.*

LANG (Andrew)—Works by—*contd.*

Grass of Parnassus. A Volume of Selected Verses. Fcp. 8vo. 6*s.*

Letters on Literature. Crown 8vo. 6*s.* 6*d.*

Old Friends: Essays in Epistolary Parody. 6*s.* 6*d.*

Ballads of Books. Edited by ANDREW LANG. Fcp. 8vo. 6*s.*

The Blue Fairy Book. Edited by ANDREW LANG. With 8 Plates and 130 Illustrations in the Text by H. J. Ford and G. P. Jacomb Hood. Crown 8vo. 6*s.*

The Red Fairy Book. Edited by ANDREW LANG. With 4 Plates and 96 Illustrations in the Text by H. J. Ford and Lancelot Speed. Crown 8vo. 6*s.*

LAVIGERIE.—**Cardinal Lavigerie and the African Slave Trade.** 1 vol. 8vo. 14*s.*

LAYARD.—**Poems.** By NINA F. LAYARD. Crown 8vo. 6*s.*

LECKY (W. E. H.)—Works by.

History of England in the Eighteenth Century. 8vo. Vols. I. & II. 1700–1760. 36*s.* Vols. III. & IV. 1760–1784. 36*s.* Vols. V. & VI. 1784–1793. 36*s.* Vols. VII. & VIII. 1793–1800. 36*s.*

The History of European Morals from Augustus to Charlemagne. 2 vols. crown 8vo. 16*s.*

History of the Rise and Influence of the Spirit of Rationalism in Europe. 2 vols. crown 8vo. 16*s.*

LEES and CLUTTERBUCK. — **B. C. 1887, A Ramble in British Columbia.** By J. A. LEES and W. J. CLUTTERBUCK. With Map and 75 Illustrations. Crown 8vo. 6*s.*

LEGER.—**A History of Austro-Hungary.** From the Earliest Time to the year 1889. By LOUIS LEGER. Translated from the French by Mrs. BIRKBECK HILL. With a Preface by E. A. FREEMAN, D.C.L. Crown 8vo. 10*s.* 6*d.*

LEWES.—**The History of Philosophy,** from Thales to Comte. By GEORGE HENRY LEWES. 2 vols. 8vo. 32*s.*

LIDDELL.—Memoirs of the Tenth Royal Hussars : Historical and Social. By Colonel LIDDELL. With Portraits and Coloured Illustration. 2 vols. Imperial 8vo.

LLOYD.—The Science of Agriculture. By F. J. LLOYD. 8vo. 12*s.*

LONGMAN (Frederick W.)—Works by.

Chess Openings. Fcp. 8vo. 2*s.* 6*d.*

Frederick the Great and the Seven Years' War. Fcp. 8vo. 2*s.* 6*d.*

Longman's Magazine. Published Monthly. Price Sixpence. Vols. 1–16, 8vo. price 5*s.* each.

Longmans' New Atlas. Political and Physical. For the Use of Schools and Private Persons. Consisting of 40 Quarto and 16 Octavo Maps and Diagrams, and 16 Plates of Views. Edited by GEO. G. CHISHOLM, M.A. B.Sc. Imp. 4to. or imp. 8vo. 12*s.* 6*d.*

LOUDON (J. C.)—Works by.

Encyclopædia of Gardening. With 1,000 Woodcuts. 8vo. 21*s.*

Encyclopædia of Agriculture; the Laying-out, Improvement, and Management of Landed Property. With 1,100 Woodcuts. 8vo. 21*s.*

Encyclopædia of Plants; the Specific Character, &c. of all Plants found in Great Britain. With 12,000 Woodcuts. 8vo. 42*s.*

LUBBOCK.—The Origin of Civilisation and the Primitive Condition of Man. By Sir J. LUBBOCK, Bart. M.P. With 5 Plates and 20 Illustrations in the text. 8vo. 18*s.*

LYALL.—The Autobiography of a Slander. By EDNA LYALL, Author of 'Donovan,' &c. Fcp. 8vo. 1*s.* sewed.

LYDE.—An Introduction to Ancient History: being a Sketch of the History of Egypt, Mesopotamia, Greece, and Rome. With a Chapter on the Development of the Roman Empire into the Powers of Modern Europe. By LIONEL W. LYDE, M.A. With 3 Coloured Maps. Crown 8vo. 3*s.*

MACAULAY (Lord).—Works of.

Complete Works of Lord Macaulay.
Library Edition, 8 vols. 8vo. £5. 5*s.*
Cabinet Edition, 16 vols. post 8vo. £4. 16*s.*

History of England from the Accession of James the Second.
Popular Edition, 2 vols. crown 8vo. 5*s.*
Student's Edition, 2 vols. crown 8vo. 12*s.*
People's Edition, 4 vols. crown 8vo. 16*s.*
Cabinet Edition, 8 vols. post 8vo. 48*s.*
Library Edition, 5 vols. 8vo. £4.

Critical and Historical Essays, with Lays of Ancient Rome, in 1 volume :
Popular Edition, crown 8vo. 2*s.* 6*d.*
Authorised Edition, crown 8vo. 2*s.* 6*d.* or 3*s.* 6*d.* gilt edges.

Critical and Historical Essays:
Student's Edition, 1 vol. crown 8vo. 6*s.*
People's Edition, 2 vols. crown 8vo. 8*s.*
Trevelyan Edition, 2 vols. crown 8vo. 9*s.*
Cabinet Edition, 4 vols. post 8vo. 24*s.*
Library Edition, 3 vols. 8vo. 36*s.*

Essays which may be had separately price 6*d.* each sewed, 1*s.* each cloth :
Addison and Walpole.
Frederick the Great.
Croker's Boswell's Johnson.
Hallam's Constitutional History.
Warren Hastings. (3*d.* sewed, 6*d.* cloth.
The Earl of Chatham (Two Essays).
Ranke and Gladstone.
Milton and Machiavelli.
Lord Bacon.
Lord Clive.
Lord Byron, and The Comic Dramatists of the Restoration.

The Essay on Warren Hastings annotated by S. HALES, 1*s.* 6*d.*
The Essay on Lord Clive annotated by H. COURTHOPE BOWEN, M.A. 2*s.* 6*d.*

Speeches :
People's Edition, crown 8vo. 3*s.* 6*d.*

Lays of Ancient Rome, &c.
Illustrated by G. Scharf, fcp. 4to. 10*s.* 6*d.*
————————— Bijou Edition, 18mo. 2*s.* 6*d.* gilt top.
————————— Popular Edition, fcp. 4to. 6*d.* sewed, 1*s.* cloth.
Illustrated by J. R. Weguelin, crown 8vo. 3*s.* 6*d.* cloth extra, gilt edges.
Cabinet Edition, post 8vo. 3*s.* 6*d.*
Annotated Edit. fcp. 8vo. 1*s.* sewed, 1*s.* 6*d.* cl.

MACAULAY (Lord)—Works of—
continued.

Miscellaneous Writings:
People's Edition, 1 vol. crown 8vo. 4s. 6d.
Library Edition, 2 vols. 8vo. 21s.

Miscellaneous Writings and Speeches:
Popular edition, 1 vol. crown 8vo. 2s. 6d.
Student's Edition, in 1 vol. crown 8vo. 6s.
Cabinet Edition, including Indian Penal Code, Lays of Ancient Rome, and Miscellaneous Poems, 4 vols. post 8vo. 24s.

Selections from the Writings of Lord Macaulay. Edited, with Occasional Notes, by the Right Hon. Sir G. O. TREVELYAN, Bart. Crown 8vo. 6s.

The Life and Letters of Lord Macaulay. By the Right Hon. Sir G. O. TREVELYAN, Bart.
Popular Edition, 1 vol. crown 8vo. 2s. 6d.
Student's Edition, 1 vol. crown 8vo. 6s.
Cabinet Edition, 2 vols. post 8vo. 12s.
Library Edition, 2 vols. 8vo. 36s.

MACDONALD (Geo.)—Works by.

Unspoken Sermons. Three Series. Crown 8vo. 3s. 6d. each.

The Miracles of Our Lord. Crown 8vo. 3s. 6d.

A Book of Strife, in the Form of the Diary of an Old Soul: Poems. 12mo. 6s.

MACFARREN—Lectures on Harmony. By Sir G. A. MACFARREN. 8vo. 12s.

MACKAIL.—Select Epigrams from the Greek Anthology. Edited, with a Revised Text, Introduction, Translation, and Notes, by J. W. MACKAIL, M.A. Fellow of Balliol College, Oxford. 8vo. 16s.

MACLEOD (Henry D.)—Works by.

The Elements of Banking. Crown 8vo. 5s.

The Theory and Practice of Banking. Vol. I. 8vo. 12s. Vol. II. 14s.

The Theory of Credit. 8vo.
Vol. I. 7s. 6d.; Vol. II. Part I. 4s. 6d. ; Vol. II. Part II. 10s. 6d.

McCULLOCH—The Dictionary of Commerce and Commercial Navigation of the late J. R. MCCULLOCH. 8vo. with 11 Maps and 30 Charts, 63s.

MALMESBURY. — Memoirs of an Ex-Minister. By the Earl of MALMESBURY. Crown 8vo. 7s. 6d.

MANUALS OF CATHOLIC PHILOSOPHY (*Stonyhurst Series*):

Logic. By RICHARD F. CLARKE, S.J. Crown 8vo. 5s.

First Principles of Knowledge. By JOHN RICKABY, S.J. Crown 8vo. 5s.

Moral Philosophy (Ethics and Natural Law). By JOSEPH RICKABY, S.J. Crown 8vo. 5s.

General Metaphysics. By JOHN RICKABY, S.J. Crown 8vo. 5s.

Psychology. By MICHAEL MAHER, S.J. Crown 8vo. 6s. 6d.

Natural Theology. By BERNARD BOEDDER, S.J. Crown 8vo. 6s. 6d.
[*Nearly ready.*

A Manual of Political Economy. By C. S. DEVAS, Esq. M.A. Examiner in Political Economy in the Royal University of Ireland. 6s. 6d. [*In preparation.*

MARTINEAU (James)—Works by.

Hours of Thought on Sacred Things. Two Volumes of Sermons. 2 vols. crown 8vo. 7s. 6d. each.

Endeavours after the Christian Life. Discourses. Crown 8vo. 7s. 6d.

The Seat of Authority in Religion. 8vo. 14s.

Essays, Reviews and Addresses. 4 vols. crown 8vo. 7s. 6d. each.
I. Personal : Political.
II. Ecclesiastical : Historical.
III. Theological : Philosophical.
IV. Academical : Religious.
[*In course of publication.*

MASON.—The Steps of the Sun: Daily Readings of Prose. Selected by AGNES MASON. 16mo. 3s. 6d.

MAUNDER'S TREASURIES.

Biographical Treasury. With Supplement brought down to 1889, by Rev. JAS. WOOD. Fcp. 8vo. 6s.

Treasury of Natural History; or, Popular Dictionary of Zoology. Fcp. 8vo. with 900 Woodcuts, 6s.

Treasury of Geography, Physical, Historical, Descriptive, and Political. With 7 Maps and 16 Plates. Fcp. 8vo. 9s.
[*Continued on next page.*

MAUNDER'S TREASURIES
—*continued*.

Scientific and Literary Treasury. Fcp. 8vo. 6*s*.

Historical Treasury: Outlines of Universal History, Separate Histories of all Nations. Fcp. 8vo. 6*s*.

Treasury of Knowledge and Library of Reference. Comprising an English Dictionary and Grammar, Universal Gazetteer, Classical Dictionary, Chronology, Law Dictionary, &c. Fcp. 8vo. 6*s*.

The Treasury of Bible Knowledge. By the Rev. J. Ayre, M.A. With 5 Maps, 15 Plates, and 300 Woodcuts. Fcp. 8vo. 6*s*.

The Treasury of Botany. Edited by J. Lindley, F.R.S. and T. Moore, F.L.S. With 274 Woodcuts and 20 Steel Plates. 2 vols. fcp. 8vo. 12*s*.

MAX MÜLLER (F.)—Works by.

Selected Essays on Language, Mythology and Religion. 2 vols. crown 8vo. 16*s*.

Lectures on the Science of Language. 2 vols. crown 8vo. 16*s*.

Hibbert Lectures on the Origin and Growth of Religion, as illustrated by the Religions of India. Crown 8vo. 7*s*. 6*d*.

Introduction to the Science of Religion; Four Lectures delivered at the Royal Institution. Crown 8vo. 7*s*. 6*d*.

Natural Religion. The Gifford Lectures, delivered before the University of Glasgow in 1888. Crown 8vo. 10*s*. 6*d*.

Physical Religion. The Gifford Lectures, delivered before the University of Glasgow in 1890. Crown 8vo. 10*s*. 6*d*.

The Science of Thought. 8vo. 21*s*.

Three Introductory Lectures on the Science of Thought. 8vo. 2*s*. 6*d*.

Biographies of Words, and the Home of the Aryas. Cr 8vo. 7*s*.6*d*

A Sanskrit Grammar for Beginners. New and Abridged Edition. By A. A. Macdonell. Crown 8vo. 6*s*.

MAY.—The Constitutional History of England since the Accession of George III. 1760-1870. By the Right Hon. Sir Thomas Erskine May, K.C.B. 3 vols. crown 8vo. 18*s*.

MEADE (L. T.)—Works by.

The O'Donnells of Inchfawn. With Frontispiece by A. Chasemore. Crown 8vo. 6*s*.

Daddy's Boy. With Illustrations. Crown 8vo. 5*s*.

Deb and the Duchess. With Illustrations by M. E. Edwards. Crown 8vo. 5*s*.

House of Surprises. With Illustrations by Edith M. Scannell. Crown 8vo. 3*s*. 6*d*.

The Beresford Prize. With Illustrations by M. E. Edwards. Crown 8vo. 5*s*.

MEATH (The Earl of)—Works by.

Social Arrows: Reprinted Articles on various Social Subjects. Cr. 8vo. 5*s*.

Prosperity or Pauperism? Physical, Industrial, and Technical Training. (Edited by the Earl of Meath). 8vo. 5*s*.

MELVILLE (G. J. Whyte)—Novels by. Crown 8vo. 1*s*. each, boards; 1*s*. 6*d*. each, cloth.

The Gladiators.	Holmby House.
The Interpreter.	Kate Coventry.
Good for Nothing.	Digby Grand.
The Queen's Maries.	General Bounce.

MENDELSSOHN.—The Letters of Felix Mendelssohn. Translated by Lady Wallace. 2 vols. cr. 8vo. 10*s*.

MERIVALE (The Very Rev. Chas.)—Works by.

History of the Romans under the Empire. Cabinet Edition, 8 vols. crown 8vo. 48*s*.
Popular Edition, 8 vols. crown 8vo. 3*s*. 6*d*. each.

The Fall of the Roman Republic: a Short History of the Last Century of the Commonwealth. 12mo. 7*s*. 6*d*.

General History of Rome from B.C. 753 to A.D. 476. Cr. 8vo. 7*s*. 6*d*.

MERIVALE (The Very Rev. Chas.)—Works by—*continued.*

The Roman Triumvirates. With Maps. Fcp. 8vo. 2s. 6d.

MILES.—The Correspondence of William Augustus Miles on the French Revolution, 1789-1817. Edited by the Rev. CHARLES POPHAM MILES, M.A. F.L.S. Honorary Canon of Durham, Membre de la Société d'Histoire Diplomatique. 2 vols. 8vo. 32s.

MILL.—Analysis of the Phenomena of the Human Mind. By JAMES MILL. 2 vols. 8vo. 28s.

MILL (John Stuart)—Works by.

Principles of Political Economy. Library Edition, 2 vols. 8vo. 30s. People's Edition, 1 vol. crown 8vo. 5s.

A System of Logic. Cr. 8vo. 5s.

On Liberty. Crown 8vo. 1s. 4d.

On Representative Government. Crown 8vo. 2s.

Utilitarianism. 8vo. 5s.

Examination of Sir William Hamilton's Philosophy. 8vo. 16s.

Nature, the Utility of Religion, and Theism. Three Essays. 8vo. 5s.

MOLESWORTH (Mrs.)—Works by.

Marrying and Giving in Marriage: a Novel. By Mrs. MOLESWORTH. Fcp. 8vo. 2s. 6d.

Silverthorns. With Illustrations by F. NOEL PATON. Crown 8vo. 5s.

The Palace in the Garden. With Illustrations by HARRIET M. BENNETT. Crown 8vo. 5s.

The Third Miss St. Quentin. Crown 8vo. 6s.

Neighbours. With Illustrations by M. ELLEN EDWARDS. Crown 8vo. 6s.

The Story of a Spring Morning, &c. With Illustrations by M. ELLEN EDWARDS. Crown 8vo. 5s.

MOON (G. Washington)—Works by.

The King's English. Fcp. 8vo. 3s. 6d.

The Soul's Inquiries Answered in the Words of Scripture. A Year-Book of Scripture Texts. Pocket Edition. Royal 32mo. 2s. 6d. Common Edition. Royal 32mo. 8d. limp; 1s. 6d. cloth.

The Soul's Desires Breathed to God in the Words of Scripture : being Prayers, and a Treatise on Prayer in the Language of the Bible. Royal 32mo. 2s. 6d.

MOORE.—Dante and his Early Biographers. By EDWARD MOORE, D.D. Principal of St. Edmund Hall, Oxford. Crown 8vo. 4s. 6d.

MULHALL.— History of Prices since the Year 1850. By MICHAEL G. MULHALL. Crown 8vo. 6s.

MURDOCK.—The Reconstruction of Europe : a Sketch of the Diplomatic and Military History of Continental Europe, from the Rise to the Fall of the Second French Empire. By HENRY MURDOCK. Crown 8vo. 9s.

MURRAY.—A Dangerous Catspaw : a Story. By DAVID CHRISTIE MURRAY and HENRY MURRAY. Cr. 8vo. 2s. 6d.

MURRAY and HERMAN. — Wild Darrie: a Story. By CHRISTIE MURRAY and HENRY HERMAN. Crown 8vo. 2s. boards ; 2s. 6d. cloth.

NANSEN.—The First Crossing of Greenland. By Dr. FRIDTJOF NANSEN. With 5 Maps, 12 Plates, and 150 Illustrations in the Text. 2 vols. 8vo. 36s.

NAPIER.—The Life of Sir Joseph Napier, Bart. Ex-Lord Chancellor of Ireland. By ALEX. CHARLES EWALD, F.S.A. With Portrait. 8vo. 15s.

NAPIER.—The Lectures, Essays, and Letters of the Right Hon. Sir Joseph Napier, Bart. late Lord Chancellor of Ireland. 8vo. 12s. 6d.

NESBIT—Leaves of Life: Verses.
By E. Nesbit. Crown 8vo. 5s.

NEWMAN.—The Letters and Correspondence of John Henry Newman during his Life in the English Church. With a brief Autobiographical Memoir. Arranged and Edited, at Cardinal Newman's request, by Miss Anne Mozley, Editor of the 'Letters of the Rev. J. B. Mozley, D.D.' With Portraits, 2 vols. 8vo. 30s. net.

NEWMAN (Cardinal)—Works by.

Apologia pro Vitâ Sua. Cabinet Edition, cr. 8vo. 6s. Cheap Edition, 3s. 6d.

Sermons to Mixed Congregations. Crown 8vo. 6s.

Occasional Sermons. Crown 8vo. 6s.

The Idea of a University defined and illustrated. Crown 8vo. 7s.

Historical Sketches. 3 vols. crown 8vo. 6s. each.

The Arians of the Fourth Century. Cabinet Edition, crown 8vo. 6s. Cheap Edition, crown 8vo. 3s. 6d.

Select Treatises of St. Athanasius in Controversy with the Arians. Freely Translated. 2 vols. cr. 8vo. 15s.

Discussions and Arguments on Various Subjects. Cabinet Edition, crown 8vo. 6s. Cheap Edition, crown 8vo. 3s. 6d.

An Essay on the Development of Christian Doctrine. Cabinet Edition, crown 8vo. 6s. Cheap Edition, crown 8vo. 3s. 6d.

Certain Difficulties felt by Anglicans in Catholic Teaching Considered. Vol. 1, crown 8vo. 7s. 6d.; Vol. 2, crown 8vo. 5s. 6d.

The Via Media of the Anglican Church, illustrated in Lectures, &c. 2 vols. crown 8vo. 6s. each.

Essays, Critical and Historical. Cabinet Edition, 2 vols. crown 8vo. 12s. Cheap Edition, 2 vols. crown 8vo. 7s.

Essays on Biblical and on Ecclesiastical Miracles. Cabinet Edition, crown 8vo. 6s. Cheap Edition, crown 8vo. 3s. 6d.

NEWMAN (Cardinal)—Works by—continued.

Tracts. 1. Dissertatiunculæ. 2. On the Text of the Seven Epistles of St. Ignatius. 3. Doctrinal Causes of Arianism. 4. Apollinarianism. 5. St. Cyril's Formula. 6. Ordo de Tempore. 7. Douay Version of Scripture. Crown 8vo. 8s.

An Essay in Aid of a Grammar of Assent. Cabinet Edition, crown 8vo. 7s. 6d. Cheap Edition, crown 8vo. 3s. 6d.

Present Position of Catholics in England. Crown 8vo. 7s. 6d.

Callista: a Tale of the Third Century. Cabinet Edition, crown 8vo. 6s. Cheap Edition, crown 8vo. 3s. 6d.

Loss and Gain: a Tale. Crown 8vo. 6s.

The Dream of Gerontius. 16mo. 6d. sewed, 1s. cloth.

Verses on Various Occasions. Cabinet Edition, crown 8vo. 6s. Cheap Edition, crown 8vo. 3s. 6d.

⁎ For Cardinal Newman's other Works see Messrs. Longmans & Co.'s Catalogue of Theological Works.

NORRIS.—Mrs. Fenton: a Sketch. By W. E. Norris. Crown 8vo. 6s.

NORTON (Charles L.)—Works by.

Political Americanisms: a Glossary of Terms and Phrases Current at Different Periods in American Politics.

A Handbook of Florida. With 49 Maps and Plans. Fcp. 8vo. 5s.

NORTHCOTT.—Lathes and Turning, Simple, Mechanical, and Ornamental. By W. H. Northcott. With 338 Illustrations. 8vo. 18s.

O'BRIEN.—When we were Boys: a Novel. By William O'Brien, M.P. Cabinet Edition, crown 8vo. 6s. Cheap Edition, crown 8vo. 2s. 6d.

OLIPHANT (Mrs.)—Novels by.

Madam. Cr. 8vo. 1s. bds.; 1s. 6d. cl.

In Trust. Cr. 8vo. 1s. bds.; 1s. 6d. cl.

Lady Car: the Sequel of a Life. Crown 8vo. 2s. 6d.

OMAN.—A History of Greece from the Earliest Times to the Macedonian Conquest. By C. W. C. OMAN, M.A. F.S.A. Fellow of All Souls College, and Lecturer at New College, Oxford. With Maps and Plans. Crown 8vo. 4*s.* 6*d.*

O'REILLY.—Hurstleigh Dene : a Tale. By Mrs. O'REILLY. Illustrated by M. ELLEN EDWARDS. Crown 8vo. 5*s.*

PAYN (James)—Novels by.

The Luck of the Darrells. Cr. 8vo. 1*s.* boards ; 1*s.* 6*d.* cloth.

Thicker than Water. Crown 8vo. 1*s.* boards ; 1*s.* 6*d.* cloth.

PERRING (Sir PHILIP)—Works by.

Hard Knots in Shakespeare. 8vo. 7*s.* 6*d.*

The 'Works and Days' of Moses. Crown 8vo. 3*s.* 6*d.*

PHILLIPPS-WOLLEY.—Snap : a Legend of the Lone Mountain. By C. PHILLIPPS-WOLLEY, Author of 'Sport in the Crimea and Caucasus' &c. With 13 Illustrations by H. G. WILLINK. Crown 8vo. 6*s.*

POLE.—The Theory of the Modern Scientific Game of Whist. By W. POLE, F.R.S. Fcp. 8vo. 2*s.* 6*d.*

POLLOCK.— The Seal of Fate : a Novel. By W. H. POLLOCK and Lady POLLOCK. Crown 8vo.

PRENDERGAST.—Ireland, from the Restoration to the Revolution, 1660-1690. By JOHN P. PRENDERGAST. 8vo. 5*s.*

PRINSEP.—Virginie : a Tale of One Hundred Years Ago. By VAL PRINSEP, A.R.A. 3 vols. crown 8vo. 25*s.* 6*d.*

PROCTOR (R. A.)—Works by.

Old and New Astronomy. 12 Parts, 2*s.* 6*d.* each. Supplementary Section, 1*s.* Complete in 1 vol. 4to. 36*s.* [*In course of publication.*]

The Orbs Around Us ; a Series of Essays on the Moon and Planets, Meteors and Comets. With Chart and Diagrams. Crown 8vo. 5*s.*

PROCTOR (R. A.)—Works by—*cont.*

Other Worlds than Ours ; The Plurality of Worlds Studied under the Light of Recent Scientific Researches. With 14 Illustrations. Crown 8vo. 5*s.*

The Moon ; her Motions, Aspects, Scenery, and Physical Condition. With Plates, Charts, Woodcuts, &c. Cr. 8vo. 5*s.*

Universe of Stars ; Presenting Researches into and New Views respecting the Constitution of the Heavens. With 22 Charts and 22 Diagrams. 8vo. 10*s.* 6*d.*

Larger Star Atlas for the Library, in 12 Circular Maps, with Introduction and 2 Index Pages. Folio, 15*s.* or Maps only, 12*s.* 6*d.*

The Student's Atlas. In Twelve Circular Maps on a Uniform Projection and one Scale. 8vo. 5*s.*

New Star Atlas for the Library, the School, and the Observatory, in 12 Circular Maps. Crown 8vo. 5*s.*

Light Science for Leisure Hours ; Familiar Essays on Scientific Subjects. 3 vols. crown 8vo. 5*s.* each.

Chance and Luck ; a Discussion of the Laws of Luck, Coincidences, Wagers, Lotteries, and the Fallacies of Gambling &c. Crown 8vo. 2*s.* boards ; 2*s.* 6*d.* cloth.

Studies of Venus-Transits. With 7 Diagrams and 10 Plates. 8vo. 5*s.*

How to Play Whist : with the Laws and Etiquette of Whist. Crown 8vo. 3*s.* 6*d.*

ome Whist : an Easy Guide to Correct Play. 16mo 1*s.*

The Stars in their Seasons. An Easy Guide to a Knowledge of the Star Groups, in 12 Maps. Roy. 8vo. 5*s.*

Star Primer. Showing the Starry Sky Week by Week, in 24 Hourly Maps. Crown 4to. 2*s.* 6*d.*

The Seasons Pictured in 48 Sun-Views of the Earth, and 24 Zodiacal Maps, &c. Demy 4to. 5*s.*

Strength and Happiness. With 9 Illustrations. Crown 8vo. 5*s.*

[*Continued on next page.*]

PROCTOR (R. A.)—Works by—*cont.*

Strength : How to get Strong and keep Strong, with Chapters on Rowing and Swimming, Fat, Age, and the Waist. With 9 Illustrations. Crown 8vo. 2*s.*

Rough Ways Made Smooth. Familiar Essays on Scientific Subjects. Crown 8vo. 5*s.*

Our Place Among Infinities. A Series of Essays contrasting our Little Abode in Space and Time with the Infinities Around us. Crown 8vo. 5*s.*

The Expanse of Heaven. Essays on the Wonders of the Firmament. Crown 8vo. 5*s.*

The Great Pyramid, Observatory, Tomb, and Temple. With Illustrations. Crown 8vo. 5*s.*

Pleasant Ways in Science. Crown 8vo. 5*s.*

Myths and Marvels of Astronomy. Crown 8vo. 5*s.*

Nature Studies. By GRANT ALLEN, A. WILSON, T. FOSTER, E. CLODD, and R. A. PROCTOR. Crown 8vo. 5*s.*

Leisure Readings. By E. CLODD, A. WILSON, T. FOSTER, A. C. RANYARD, and R. A. PROCTOR. Crown 8vo. 5*s.*

PRYCE.—The Ancient British Church : an Historical Essay. By JOHN PRYCE, M. A. Crown 8vo. 6*s.*

RANSOME.—The Rise of Constitutional Government in England: being a Series of Twenty Lectures on the History of the English Constitution delivered to a Popular Audience. By CYRIL RANSOME, M.A. Crown 8vo. 6*s.*

RAWLINSON.—The History of Phœnicia. By GEORGE RAWLINSON, M.A. Canon of Canterbury, &c. With numerous Illustrations. 8vo. 24*s.*

READER.—Echoes . of Thought : a Medley of Verse. By EMILY E. READER. Fcp. 8vo. 5*s.* cloth, gilt top.

RENDLE and NORMAN.—The Inns of Old Southwark, and their Associations. By WILLIAM RENDLE, F.R.C.S. and PHILIP NORMAN, F.S.A. With numerous Illustrations. Roy.8vo.28*s.*

RIBOT.—The Psychology of Attention. By TH. RIBOT. Crown 8vo. 3*s.*

RICH.—A Dictionary of Roman and Greek Antiquities. With 2,000 Woodcuts. By A. RICH. Cr. 8vo. 7*s. 6d.*

RICHARDSON.—National Health. Abridged from 'The Health of Nations.' A Review of the Works of Sir Edwin Chadwick, K.C.B. By Dr. B. W. RICHARDSON. Crown, 4*s. 6d.*

RILEY.—Athos ; or, the Mountain of the Monks. By ATHELSTAN RILEY, M.A. F.R.G.S. With Map and 29 Illustrations. 8vo. 21*s.*

RIVERS. — The Miniature Fruit Garden ; or, the Culture of Pyramidal and Bush Fruit Trees. By THOMAS RIVERS. With 32 Illustrations. Fcp. 8vo. 4*s.*

ROBERTS.—Greek the Language of Christ and His Apostles. By ALEXANDER ROBERTS, D.D. 8vo. 18*s.*

ROGET.—A History of the 'Old Water-Colour' Society (now the Royal Society of Painters in Water-Colours). With Biographical Notices of its Older and all its Deceased Members and Associates. Preceded by an Account of English Water-Colour Art and Artists in the Eighteenth Century. By JOHN LEWIS ROGET, M.A. Barrister-at-Law. 2 vols. royal 8vo.

ROGET.—Thesaurus of English Words and Phrases. Classified and Arranged so as to facilitate the Expression of Ideas. By PETER M. ROGET. Crown 8vo. 10*s. 6d*

RONALDS.—The Fly-Fisher's Entomology. By ALFRED RONALDS. With 20 Coloured Plates. 8vo. 14*s.*

ROSSETTI.—A Shadow of Dante: being an Essay towards studying Himself, his World, and his Pilgrimage. By MARIA FRANCESCA ROSSETTI. With Illustrations. Crown 8vo. 10s. 6d.

RUSSELL.—A Life of Lord John Russell (Earl Russell, K.G.). By SPENCER WALPOLE. With 2 Portraits. 2 vols. 8vo. 36s. Cabinet Edition, 2 vols. crown 8vo. 12s.

SEEBOHM (Frederic)—Works by.

The Oxford Reformers—John Colet, Erasmus, and Thomas More; a History of their Fellow-Work. 8vo. 14s.

The Era of the Protestant Revolution. With Map. Fcp. 8vo. 2s. 6d.

The English Village Community Examined in its Relations to the Manorial and Tribal Systems, &c, 13 Maps and Plates. 8vo. 16s.

SEWELL.—Stories and Tales. By ELIZABETH M. SEWELL. Crown 8vo. 1s. 6d. each, cloth plain; 2s. 6d. each, cloth extra, gilt edges:—

Amy Herbert.	Laneton Parsonage.
The Earl's Daughter.	Ursula.
The Experience of Life.	Gertrude.
A Glimpse of the World.	Ivors.
Cleve Hall.	Home Life.
Katharine Ashton.	After Life.
Margaret Percival.	

SHAKESPEARE.—Bowdler's Family Shakespeare. 1 vol. 8vo. With 36 Woodcuts, 14s. or in 6 vols. fcp. 8vo. 21s.

Outlines of the Life of Shakespeare. By J. O. HALLIWELL-PHILLIPPS. 2 vols. Royal 8vo. £1. 1s.

Shakespeare's True Life. By JAMES WALTER. With 500 Illustrations. Imp. 8vo. 21s.

The Shakespeare Birthday Book. By MARY F. DUNBAR. 32mo. 1s. 6d. cloth. With Photographs, 32mo. 5s. Drawing-Room Edition, with Photographs, fcp. 8vo. 10s. 6d.

SHORT.—Sketch of the History of the Church of England to the Revolution of 1688. By T. V. SHORT, D.D. Crown 8vo. 7s. 6d.

SMITH (Gregory).—Fra Angelico, and other Short Poems. By GREGORY SMITH. Crown 8vo. 4s. 6d.

SMITH (R. Bosworth).—Carthage and the Carthagenians. By R. BOSWORTH SMITH, M.A. Maps, Plans, &c. Crown 8vo. 6s.

Sophocles. Translated into English Verse. By ROBERT WHITELAW, M.A. Assistant-Master in Rugby School; late Fellow of Trinity College, Cambridge. Crown 8vo. 8s. 6d.

STANLEY.—A Familiar History of Birds. By E. STANLEY, D.D. With 160 Woodcuts. Crown 8vo. 3s. 6d.

STEEL (J. H.)—Works by.

A Treatise on the Diseases of the Dog; being a Manual of Canine Pathology. Especially adapted for the Use of Veterinary Practitioners and Students. 88 Illustrations. 8vo. 10s. 6d.

A Treatise on the Diseases of the Ox; being a Manual of Bovine Pathology specially adapted for the use of Veterinary Practitioners and Students. 2 Plates and 117 Woodcuts. 8vo. 15s.

A Treatise on the Diseases of the Sheep: being a Manual of Ovine Pathology. Especially adapted for the use of Veterinary Practitioners and Students. With Coloured Plate and 99 Woodcuts. 8vo. 12s.

STEPHEN. — Essays in Ecclesiastical Biography. By the Right Hon. Sir J. STEPHEN. Cr. 8vo. 7s. 6d.

STEPHENS.—A History of the French Revolution. By H. MORSE STEPHENS, Balliol College, Oxford. 3 vols. 8vo. Vol. I. 18s. *Ready.* *Vol. II. in the press.*

STEVENSON (Robt. Louis)—Works by.

A Child's Garden of Verses. Small fcp. 8vo. 5s.

The Dynamiter. Fcp. 8vo. 1s. swd. 1s. 6d. cloth.

Strange Case of Dr. Jekyll and Mr. Hyde. Fcp. 8vo. 1s. swd.; 1s. 6d. cloth.

STEVENSON and OSBOURNE.—The Wrong Box. By ROBERT LOUIS STEVENSON and LLOYD OSBOURNE. Crown 8vo. 5s.

STOCK.—Deductive Logic. By ST. GEORGE STOCK. Fcp. 8vo. 3s. 6d.

'STONEHENGE.'—The Dog in Health and Disease. By 'STONEHENGE.' With 84 Wood Engravings. Square crown 8vo. 7s. 6d.

STRONG and LOGEMAN.—Introduction to the Study of the History of Language. By HERBERT A. STRONG, M.A. LL.D.; WILLEM S. LOGEMAN; and BENJAMIN IDE WHEELER. 8vo. 10s. 6d.

SULLY (James)—Works by.

Outlines of Psychology, with Special Reference to the Theory of Education. 8vo. 12s. 6d.

The Teacher's Handbook of Psychology, on the Basis of 'Outlines of Psychology.' Cr. 8vo. 6s. 6d.

Supernatural Religion ; an Inquiry into the Reality of Divine Revelation. 3 vols. 8vo. 36s.

Reply (A) to Dr. Lightfoot's Essays. By the Author of 'Supernatural Religion.' 1 vol. 8vo. 6s.

SWINBURNE.—Picture Logic ; an Attempt to Popularise the Science of Reasoning. By A. J. SWINBURNE, B.A. Post 8vo. 5s.

SYMES.—Prelude to Modern History : being a Brief Sketch of the World's History from the Third to the Ninth Century. By J. E. SYMES, M.A. University College, Nottingham. With 5 Maps. Crown 8vo. 2s. 6d.

TAYLOR.—A Student's Manual of the History of India, from the Earliest Period to the Present Time. By Colonel MEADOWS TAYLOR, C.S.I. &c. Crown 8vo. 7s. 6d.

THOMPSON (D. Greenleaf)—Works by.

The Problem of Evil : an Introduction to the Practical Sciences. 8vo. 10s. 6d.

THOMPSON (D. Greenleaf)—Works by—*continued.*

A System of Psychology. 2 vols. 8vo. 36s.

The Religious Sentiments of the Human Mind. 8vo. 7s. 6d.

Social Progress: an Essay. 8vo. 7s. 6d.

The Philosophy of Fiction in Literature : an Essay. Cr. 8vo. 6s.

Three in Norway. By Two of THEM. With a Map and 59 Illustrations. Cr. 8vo. 2s. boards; 2s. 6d. cloth.

TOYNBEE.—Lectures on the Industrial Revolution of the 18th Century in England. By the late ARNOLD TOYNBEE, Tutor of Balliol College, Oxford. Together with a Short Memoir by B. JOWETT, Master of Balliol College, Oxford. 8vo. 10s. 6d.

TREVELYAN (Sir G. O. Bart.)—Works by.

The Life and Letters of Lord Macaulay.

POPULAR EDITION, 1 vol. cr. 8vo. 2s. 6d.

STUDENT'S EDITION, 1 vol. cr. 8vo. 6s.

CABINET EDITION, 2 vols. cr. 8vo. 12s.

LIBRARY EDITION, 2 vols. 8vo. 36s.

The Early History of Charles James Fox. Library Edition, 8vo. 18s. Cabinet Edition, crown 8vo. 6s.

TROLLOPE (Anthony).—Novels by.

The Warden. Crown 8vo. 1s. boards; 1s. 6d. cloth.

Barchester Towers. Crown 8vo. 1s. boards ; 1s. 6d. cloth.

VILLE.—On Artificial Manures, their Chemical Selection and Scientific Application to Agriculture. By GEORGES VILLE. Translated and edited by W. CROOKES. With 31 Plates. 8vo. 21s.

VIRGIL.—Publi Vergili Maronis Bucolica, Georgica, Æneis; the Works of VIRGIL, Latin Text, with English Commentary and Index. By B. H. KENNEDY, D.D. Cr. 8vo. 10s. 6d.

The Æneid of Virgil. Translated into English Verse. By JOHN CONINGTON, M.A. Crown 8vo. 6s.

The Poems of Virgil. Translated into English Prose. By JOHN CONINGTON, M.A. Crown 8vo. 6s.

The Eclogues and Georgics of Virgil. Translated from the Latin by J. W. MACKAIL, M.A. Fellow of Balliol College, Oxford. Printed on Dutch Hand-made Paper. Royal 16mo. 5s.

WAKEMAN and HASSALL.—Essays Introductory to the Study of English Constitutional History. By Resident Members of the University of Oxford. Edited by HENRY OFFLEY WAKEMAN, M.A. Fellow of All Souls College, and ARTHUR HASSALL, M.A. Student of Christ Church. Crown 8vo. 6s.

WALKER.—The Correct Card; or How to Play at Whist; a Whist Catechism. By Major A. CAMPBELL-WALKER, F.R.G.S. Fcp. 8vo. 2s. 6d.

WALPOLE.—History of England from the Conclusion of the Great War in 1815 to 1858. By SPENCER WALPOLE. Library Edition. ·5 vols. 8vo. £4. 10s. Cabinet Edition. 6 vols. crown 8vo. 6s. each.

WELLINGTON.—Life of the Duke of Wellington. By the Rev. G. R. GLEIG, M.A. Crown 8vo. 3s. 6d.

WELLS. — Recent Economic Changes and their Effect on the Production and Distribution of Wealth and the Well-being of Society. By DAVID A. WELLS, LL.D. D.C.L. late United States Special Commissioner of Revenue, &c. Crown 8vo. 10s. 6d.

WENDT.— Papers on Maritime Legislation, with a Translation of the German Mercantile Laws relating to Maritime Commerce. By ERNEST EMIL WENDT, D.C.L. Royal 8vo. £1. 11s. 6d.

WEST.—Lectures on the Diseases of Infancy and Childhood. By CHARLES WEST, M.D. 8vo. 18s.

WEYMAN.—The House of the Wolf: a Romance. By STANLEY J. WEYMAN. Crown 8vo. 6s.

WHATELY (E. Jane)—Works by.

English Synonyms. Edited by R. WHATELY, D.D. Fcp. 8vo. 3s.

Life and Correspondence of Richard Whately, D.D. late Archbishop of Dublin. With Portrait. Crown 8vo. 10s. 6d.

WHATELY (Archbishop)—Works by.

Elements of Logic. Cr. 8vo. 4s. 6d.

Elements of Rhetoric. Crown 8vo. 4s. 6d.

Lessons on Reasoning. Fcp. 8vo. 1s. 6d.

Bacon's Essays, with Annotations. 8vo. 10s. 6d.

WILCOCKS.—The Sea Fisherman. Comprising the Chief Methods of Hook and Line Fishing in the British and other Seas, and Remarks on Nets, Boats, and Boating. By J. C. WILCOCKS. Profusely Illustrated. Crown 8vo. 6s.

WILLICH. — Popular Tables for giving Information for ascertaining the value of Lifehold, Leasehold, and Church Property, the Public Funds, &c. By CHARLES M. WILLICH. Edited by H. BENCE JONES. Crown 8vo. 10s. 6d.

WILLOUGHBY.—East Africa and its Big Game. The Narrative of a Sporting Trip from Zanzibar to the Borders of the Masai. By Capt. Sir JOHN C. WILLOUGHBY, Bart. Illustrated by G. D. Giles and Mrs. Gordon Hake. Royal 8vo. 21s.

WITT (Prof.)—Works by. Translated by FRANCES YOUNGHUSBAND.

The Trojan War. Crown 8vo. 2s.

Myths of Hellas; or, Greek Tales. Crown 8vo. 3s. 6d.

WITT (Prof.)—Works by—*cont.*

The Wanderings of Ulysses.
Crown 8vo. 3*s.* 6*d.*

The Retreat of the Ten Thousand; being the Story of Xenophon's 'Anabasis.' With Illustrations.

WOLFF.—Rambles in the Black Forest. By HENRY W. WOLFF. Crown 8vo. 7*s.* 6*d.*

WOOD (Rev. J. G.)—Works by.

Homes Without Hands; a Description of the Habitations of Animals, classed according to the Principle of Construction. With 140 Illustrations. 8vo. 10*s.* 6*d.*

Insects at Home; a Popular Account of British Insects, their Structure, Habits, and Transformations. With 700 Illustrations. 8vo. 10*s.* 6*d.*

Insects Abroad; a Popular Account of Foreign Insects, their Structure, Habits, and Transformations. With 600 Illustrations. 8vo. 10*s.* 6*d.*

Bible Animals; a Description of every Living Creature mentioned in the Scriptures. With 112 Illustrations. 8vo. 10*s.* 6*d.*

Strange Dwellings; a Description of the Habitations of Animals, abridged from 'Homes without Hands.' With 60 Illustrations. Crown 8vo. 3*s.* 6*d.*

Out of Doors; a Selection of Original Articles on Practical Natural History. With 11 Illustrations. Crown 8vo. 3*s.* 6*d.*

Petland Revisited. With 33 Illustrations. Crown 8vo. 3*s.* 6*d.*

The following books are extracted from the foregoing works by the Rev. J. G. WOOD:

Social Habitations and Parasitic Nests. With 18 Illustrations. Crown 8vo. 2*s.*

The Branch Builders. With 28 Illustrations. Crown 8vo. 2*s.* 6*d.*

Wild Animals of the Bible. With 29 Illustrations. Crown 8vo. 3*s.* 6*d.*

Domestic Animals of the Bible. With 23 Illustrations. Crown 8vo. 3*s.* 6*d.*

WOOD (Rev. J. G.)—Works by—*cont.*

Bird-Life of the Bible. With 32 Illustrations. Crown 8vo. 3*s.* 6*d.*

Wonderful Nests. With 30 Illustrations. Crown 8vo. 3*s.* 6*d.*

Homes under the Ground. With 28 Illustrations. Crown 8vo. 3*s.* 6*d.*

YOUATT (William)—Works by.

The Horse. Revised and enlarged. 8vo. Woodcuts, 7*s.* 6*d.*

The Dog. Revised and enlarged. 8vo. Woodcuts. 6*s.*

YOUNGHUSBAND (Frances)—Works by.

The Story of our Lord, told in Simple Language for Children. With 25 Illustrations on Wood from Pictures by the Old Masters. Crown 8vo. 2*s.* 6*d.*

The Story of Genesis. Crown 8vo. 2*s.* 6*d.*

ZELLER (Dr. E.)—Works by.

History of Eclecticism in Greek Philosophy. Translated by SARAH F. ALLEYNE. Crown 8vo. 10*s.* 6*d.*

The Stoics, Epicureans, and Sceptics. Translated by the Rev. O. J. REICHEL, M.A. Crown 8vo. 15*s.*

Socrates and the Socratic Schools. Translated by the Rev. O. J. REICHEL, M.A. Crown 8vo. 10*s.* 6*d.*

Plato and the Older Academy. Translated by SARAH F. ALLEYNE and ALFRED GOODWIN, B.A. Crown 8vo. 18*s.*

The Pre-Socratic Schools: a History of Greek Philosophy from the Earliest Period to the time of Socrates. Translated by SARAH F. ALLEYNE. 2 vols. crown 8vo. 30*s.*

Outlines of the History of Greek Philosophy. Translated by SARAH F. ALLEYNE and EVELYN ABBOTT. Crown 8vo. 10*s.* 6*d.*

5,000/3/91   Spottiswoode & Co. Printers New-street Square, London.